"In this book Chemhuru ably articulates the bearing of African philosophies on environmental justice. His book will be essential reading for Africans concerned about how their traditions can motivate and underpin environmental equity, and worldwide for all who seek to discover what can be learned about relations between genders, classes, cultures and generations from the continent where humanity first evolved".

Robin Attfield, *Professor Emeritus of Philosophy at Cardiff University, Wales, UK.*

"Here comes a resourceful publication on environmental justice with convincing arguments and illustrations on why and how everyone is a stakeholder in the project and must be involved in achieving environmental justice for mankind. It deserves global attention".

Lawrence Ogbo Ugwuanyi, *Professor of Philosophy, University of Abuja, Nigeria.*

I0131861

Environmental Justice in African Philosophy

This book focuses on environmental justice in African philosophy, highlighting important new perspectives which will be of significance to researchers with an interest in environmental ethics both within Africa and beyond.

Drawing on African social and ethical conceptions of existence, the book makes suggestions for how to derive environmental justice from African philosophies such as communitarian ethics, relational ethics, unhu/ubuntu ethics, ecofeminist ethics and intergenerational ethics. Specifically, the book emphasises the ways in which African philosophies of existence seek to involve everyone in environmental policy and planning and to equitably distribute both environmental benefits (such as natural resources) and environmental burdens (such as pollution and the location of mining, industrial or dumping sites). This extends to fair distribution between global South and global North, rich and poor, urban and rural populations, men and women and adults and children. These principles of humaneness, relationships, equality, interconnectedness and teleologically oriented existence among all beings are important not only to African environmental justice but also to the environmental justice movement globally.

The book will interest researchers and students working in the fields of environmental ethics, African philosophy and political philosophy in general.

Munamato Chemhuru is an Associate Professor of Philosophy in the Department of Philosophy and Religious Studies at Great Zimbabwe University and a Senior Research Associate in Philosophy in the Faculty of Humanities at the University of Johannesburg. Munamato is a Georg Forster Research Fellow under the Alexander von Humboldt Foundation at Katholische Universität Eichstätt-Ingolstadt, Germany (2020-2022).

Routledge Studies in African Philosophy

African Philosophy and the Otherness of Albinism
White Skin, Black Race
Elvis Imafidon

Africanizing African Legal Ethics
John Murungi

Consolationism and Comparative African Philosophy
Beyond Universalism and Particularism
Ada Agada

Futurism and the African Imagination
Literature and Other Arts
Edited by Dike Okoro

Critical Conversations in African Philosophy
Asixoxe - Let's Talk
Edited by Alena Rettová, Benedetta Lanfranchi and Miriam Pahl

Environmental Justice in African Philosophy
Munamato Chemhuru

For more information about this series, please visit: https://www.routledge.com/Routledge-Studies-in-African-Philosophy/book-series/AFRPHIL

Environmental Justice in African Philosophy

Munamato Chemhuru

Routledge
Taylor & Francis Group

LONDON AND NEW YORK

First published 2022
by Routledge
4 Park Square, Milton Park, Abingdon, Oxon OX14 4RN

and by Routledge
605 Third Avenue, New York, NY 10158

Routledge is an imprint of the Taylor & Francis Group, an informa business

© 2022 Munamato Chemhuru

The right of Munamato Chemhuru to be identified as author of this work has been asserted in accordance with sections 77 and 78 of the Copyright, Designs and Patents Act 1988.

British Library Cataloguing-in-Publication Data
A catalogue record for this book is available from the British Library

Library of Congress Cataloging-in-Publication Data
A catalog record has been requested for this book

ISBN: 978-1-032-00667-3 (hbk)
ISBN: 978-1-032-00668-0 (pbk)
ISBN: 978-1-003-17671-8 (ebk)

DOI: 10.4324/9781003176718

Typeset in Bembo
by KnowledgeWorks Global Ltd.

Contents

Acknowledgements

I am indebted to many people without whose assistance I would not have been able to write this book. The writing of this book was fully funded through the Georg Forster Research Fellowship award from the Alexander von Humboldt Foundation. During my research fellowship, I was stationed in the Department of Philosophy and Systematic Pedagogics of the Faculty of Philosophy of Education at Katholische Universität Eichstätt-Ingolstadt, Eichstätt, Germany. I would like to thank Professor Dr Kai Horsthemke who hosted me as an Alexander von Humboldt fellow in the department. His academic guidance contributed considerably to the success of this book project. My gratitude also goes to members of the Department of Philosophy and Systematic Pedagogics at the Katholische Universität Eichstätt-Ingolstadt for providing me with useful feedback when I presented the ideas relating to this book at a research colloquium on 3 July 2020.

I also wish to express my appreciation to my university, Great Zimbabwe University, for granting me leave so that I could take up this fellowship between March 2020 and April 2022. I am grateful, too, to the publishers, *Routledge: Taylor and Francis* for showing confidence in my work. I am especially thankful to the Series Editors, Leanne Hinves with whom I worked from the proposal stage, and to Helena Hurd, as well as the Editorial Assistant, Rosie Anderson. Special thanks are extended to my academic colleagues, Dennis Masaka and Tendai Mangena, and the anonymous reviewers, whose constructive criticisms helped to strengthen some of the arguments and positions expressed in this book. Thanks are also due to Mrs Margret Chipara of the University of Zimbabwe who has also read each and every chapter in this book and provided me with very useful editorial corrections that have improved the quality of this book.

I also wish to thank the publishers, for allowing me to adapt and reproduce the following works, which were previously published in their *Taylor & Francis* journals:

Chapter 2, was adapted from: Munamato Chemhuru, (2019). The Paradox of Global Environmental Justice: Appealing to the Distributive Justice Framework for the Global South, *South African Journal of Philosophy*, 38 (1), pp.30–39, DOI: 10.1080/02580136.2019.1570712.

Chapter 5, was adapted from: Munamato Chemhuru, (2019). Interpreting Ecofeminist Environmentalism in African Communitarian Philosophy and *Ubuntu*: An Alternative to Anthropocentrism, *Philosophical Papers*, 48 (2), pp.241-264, DOI: 10.1080/05568641.2018.1450643.

Finally, it would be remiss of me not to acknowledge the support and love from my family, particularly my wife Anna, during what has been a very difficult two-year *sabbatical* from the family. However, she has managed to take good care of our family, especially our children; Anotidaishe, Tadiwanashe, Akudzweishe and Tavongaishe, while I was in Germany working on this book.

M.C

Eichstätt, Germany, 2022

Introduction

The idea of environmental justice cannot be separated from issues of environmental policy, planning and practice, as well as problems associated with them. At the global level, generally, the reality of climate change resulting from poor environmental policies, planning and practices is currently being experienced through loss of biodiversity, heat waves, global warming or upsurges in global temperatures, resulting from increases in 'greenhouse gases' among others. However, the effects of poor environmental policies, planning and practices all over the world, which scientists believe might be irreversible in the near future, tend to affect people in different ways.

In the global South, African communities – particularly the rural populace, disadvantaged, people with disabilities, the poor, women and children – tend to bear the burden of unjust environmental policies and practices. In these communities, dumping sites and heavy industries are for the most part, located closer to high-density residential areas, where the majority of poor people live. As former United States president, Barack Obama, observes, "environmentally problematic facilities tend to be located in places where poor folks live" (Obama, 2017). Although Obama might not have been speaking directly about the African condition, his view aptly captures the nature of the environmental justice issues faced by the majority of countries in Africa and the rest of the world. Notwithstanding as John Rawls reminds us, "justice denies that the loss of freedom of some is made right by a greater good shared by others" (Rawls, 1971: 3). This shows how the idea of justice, in spite of its long history, remains essential to contemporary society, where environmental justice issues are now more urgent than never before, especially in Africa.

Although issues around environmental justice might appear like recent, in Africa, environmental injustice may be largely attributable to the history of slavery and colonialism. Elsewhere, I have shown how slavery and colonialism, in particular, have largely contributed to imbalances in land and natural resources ownership, inequalities, poverty and, ultimately, environmental injustice (Chemhuru, 2021: 228). In most African communities, land – which for Sam Moyo is always a "basic source of livelihood" (Moyo, 2005: 146) and one of the most fundamental natural resources for safeguarding human wellbeing and property rights (Aristotle, 2011: NE: Part III-VIII) – is still

DOI: 10.4324/9781003176718-1

owned by non-indigenous populations, who constitute the minority. The indigenous people, who constitute the majority, do not have access to it. Erasmus Masitera observes that the current "land arrangement is out of touch with contemporary African interests. The interests of equitable distribution of resources burdens and benefits linked to land ownership and use" (Masitera, 2021: 2). Implicit in this observation is the environmental injustice resulting from lack of access to land and the means of production. This is even worse when mining, agriculture, industry and tourism activities are, as is the norm, owned by foreign transnational corporations and located close to rural and poor communities. As a result, indigenous local communities are always exposed to poverty, air and water pollution, human-wildlife conflict because they lack access to the means of production and live close to extractive mining and agricultural industries and tourist attractions such as national parks and conservation areas, which have little or no direct benefit for them (Kelbessa, 2009: 10). This is aggravated by the fact that the rich owners of these mines are transnational companies that are located outside Africa. Environmental justice issues in Africa are therefore quite complex. This book, which is not intended to be prescriptive, offers fresh perspectives into how African philosophy might be understood in considering environmental justice and related issues.

Ordinary African people tend to be accorded less attention in environmental justice issues. As Margaret Ssebunya, Stephen Nkansah Motgan and Beatrice D. Okyere-Manu observe, "current debates and discussions on environmental justice seem to focus more on the West" (Ssebunya, Morgan and Okyere-Manu, 2019: 175). This, despite the fact that sub-Saharan Africa has long been subjected to more environmental injustice than communities in the global North. More often than not, when scholars and policy-makers in the global North consider environmental justice and climate crisis issues, they tend to be concerned with those that affect them alone. They do not usually pay much attention to the fact that poor African communities in the global South are the most affected by these issues.

The book therefore seeks to respond to questions about what African environmental justice conceptions look like. It does so by transcending the generally accepted interpretations of African environmental ethics in order to arrive at plausible views of African environmental justice. Although African people are still obliged to show commitment to global environmental justice broadly, they could begin by searching for local approaches to local problems first, before considering how these local solutions might also contribute to environmental justice at a global scale. This approach resonates well with the quest for local knowledge production for local solutions. It also satisfies what Sabelo J. Ndlovu-Gatsheni sees as "epistemic freedom", which is necessary for the "deprovincialization" and "decolonization" of knowledge in Africa (Ndlovu-Gatsheni, 2018: 1). Although this book will focus particularly on knowledge production, it will consider how local (e.g., African) knowledge might be understood in the quest for environmental justice. It foregrounds

African philosophy as well as theoretical and practical approaches in order to address the major research question, namely how African conceptions of environmental justice could be gleaned from African philosophy.

The book is written against the background of a general lack of comprehensive, coherent and unified accounts on African environmental justice in existing literature on African environmental ethics. However, taking into account the observation that "Africa is rich with moral values and ethical traditions, which to a great extent have not been investigated by scholars" (Murove, 2020: 7), it attempts to offer some plausible approaches to African environmental justice based on various conceptions of existence in African philosophy. The book thus signals to the reader the various interpretations of the African views of existence ranging from communitarian accounts (see Mbiti, 1969; Menkiti, 1984, 2004; Gyekye, 2007, 2010, 2013; Kalumba, 2020), the philosophy of *ubuntu* (Samkange and Samkange, 1980; Ramose, 1999; Metz, 2011), as well as other emerging interpretations of African environmental ethics such as the ecofeminist view and the intergenerational environmental ethical view, all of which emphasise relational living (Metz, 2019), harmony, equality and justice amongst all beings. These interpretations of African philosophy augur well with the need for taking into account "the contribution of African ethical traditions in the day-to-day reflections of African socio-economic, political, psychological and religious realities" (Murove, 2020: 7), such as environmental justice, which is increasingly becoming an area of serious concern for both academics and policy makers.

It would be quite a difficult, if not impossible, task to undertake a general examination of environmental justice in African philosophy. *Environmental Justice in African Philosophy* therefore deals essentially with issues relating to environmental ethics and environmental justice issues in African philosophy. This book is divided into six chapters. Chapter 1, *Environmental Ethics in African Philosophy*, discusses how environmental ethics as "a theory and practice about appropriate concern for, values in, and duties towards the natural world" (Rolston, 1999: 407) could be interpreted in African philosophy. It examines African philosophical conceptions about how to ground the moral status of various aspects of nature, animal rights and welfare, global warming, climate change, poverty, pollution, and the extinction of biodiversity among other things, within the African philosophical tradition. This chapter uses a primarily historical approach, similar to Collingwood (1946)'s "philosophy of history", to trace and analyse environmental ethical thinking in African philosophy despite some prior misconceptions concerning the status of African philosophy and ultimately African environmental philosophy. The chapter broadly makes the claim that the discussion of questions relating to African environmental ethics should be preceded by a proper interpretation of African philosophy.

One of the most important environmental justice questions in African environmental ethics is how can one explain the current disparities in the distributive patterns of environmental 'goods' and 'bads' amongst communities in the global North and those in the global South? People in the global South appear to be the most adversely affected by environmental burdens and enjoy the least environmental benefits, when compared to their counterparts in the global North. Ironically, those in the global South have, through the course of history, contributed little towards the current environmental and climate change conditions (Beer, 2014: 85). For these reasons, the question of environmental justice and injustice in Africa should not just be an African affair. Yet it is an issue on which both the global North and South disagree. The notion of environmental injustice in Africa in particular (and in the global South at large) invokes other historical, social, political and economic issues which have been largely attributed to the global North. This is why "negotiations for a global agreement to address climate change have often pitted the nations of the heavily industrialised global North against the nations of the global South" (Beer, 2014: 84). It is not surprising, therefore, that environmental justice is increasingly becoming a human rights issue in Africa. It is for these reasons that Chapter 2, *Environmental in (Justice) in Africa: The North – South Challenge,* considers some of the social and political hurdles that are often encountered in the quest for environmental justice in Africa. It seeks to confront some of the historical, social, political and economic issues in order to offer a plausible distributive framework of environmental justice for Africa based on human and economic rights.

That environmental justice issues in African philosophy are essentially human rights issues cannot be disputed. This and the view that human beings *belong* to the land or to the environment at large are also not new to African philosophy (Conradie, 2019: 127). What this means is that, over and above considering the environment as being at the centre of basic human rights and justice, there should be a close ontological connection between human beings and the environment. In keeping with such a view, Chapter 3, *Environmental Justice from an African Land Ethic,* seeks to examine how an ontological understanding of human beings and the land ethic might be closely connected with human rights and environmental justice issues in African philosophy. Notwithstanding some of the metaphysical premises on which the land ethic view is based, and from which it might be objected, the chapter situates the land ethic within the African philosophical, and largely metaphysical context. It then tries to provide reasonable grounds on how to conceptualise environmental justice from it, based on what it means for human beings to be ontologically connected to the land.

The question of how far relationships could go towards inculcating a sense of duty and obligation with respect to environmental justice is also important in African relational environmental ethics. The African understanding of relationships largely stems from the idea of communitarian existence (Mbiti, 1969; Menkiti, 1984; Gyekye, 1992; Menkiti, 2004) over and above the obvious maternal and paternal relationships that exist among individuals. Although the notion of African communitarian existence remains one of the most influential views in African environmental ethics, the idea of relationships remains largely underexplored in terms of its import to specific conceptions of environmental justice. A considerable body of literature has been produced on African relational environmental ethics, analysing how far relationships can go towards promoting environmental ethics in African philosophy (Murove, 2004; Behrens, 2014; Metz, 2019; Mweshi, 2019). For Behrens, "a promising African environmentalism can be found in a belief in a fundamental interrelatedness between natural objects. What establishes moral considerability on this African view is that entities are part of the interrelated web of life" (Behrens, 2014: 63). Thus, the relational view of ethics considers the moral considerability of entities on the basis of their degree of relationship with humans (Metz, 2019). Chapter 4, *African Relational Environmental Justice,* extends beyond most views that tend to focus on the moral considerability of various beings on the basis of relationships. It attempts to discover some of the far-reaching and non-anthropocentric implications of African relational ethics to environmental justice emanating from such moral considerability on the basis of relationships between human and non-human entities.

There is a long-standing claim that a very close connection exists between social justice issues and the unjustified domination of nature (Cuomo, 2002: 1). This claim is central to understanding environmental justice issues affecting what are traditionally viewed as "human Others" (Warren, 2000: 1) such as women, black people, children, the disadvantaged, people with disabilities and the poor. It is for this reason that any ecofeminist environmental justice framework is built on an attempt to understand social justice virtues such as equality, respect, justice, interdependence and harmony for both "human Others" and "earth Others" (Warren, 2000: 1). These virtues are used in order to confront traditional dualisms and patriarchy that perpetuate inequalities and oppression responsible for social and environmental injustice. Chapter 5, *African Ecofeminist Environmental Justice,* focuses on the prospects for constructing closer connections between ecofeminist and environmental justice issues. It specifically seeks to investigate African conceptions of environmental justice using similar African ecofeminist perspectives in African environmental ethics. The justification for such an ecofeminist approach to environmental justice is that ecological problems resulting from environmental injustice such as pollution, the extinction of species and dumping of toxins mostly affect women, children, black people, the disadvantaged, people with disabilities and poor people who are, for the most part located in Africa. For this reason, the quest for African environmental justice ought to incorporate the gendered and ecofeminist analysis of similar ecological issues in African philosophy.

African environmental justice, like any conception of environmental justice, would be incomplete if it does not consider the question of how to "save justice between generations" (Rawls, 1971: 228). At the same time, it is very difficult to justify the considerability and duties towards future humans and non-humans whose existence we are not actually sure of. Nevertheless, intergenerational environmental justice issues remain at the core of any meaningful environmental justice theory. Questions of environmental justice should not be limited to present beings alone without considering the intra- and interconnectedness between and among beings and potential beings belonging to different generations. Contrary to some misconceptions about African environmental ethics, African philosophies of existence offer comprehensive perspectives about how to ground intergenerational environmental justice. Chapter 6, *Intergenerational Environmental Justice in African Philosophy,* therefore explores the intergenerational environmental justice perspective in African philosophy. It considers how African philosophies of existence might be reasonably understood in order to see how they can be used in equitably saving environmental justice between different generations, including future generations.

Each of the six chapters in this book therefore discusses different, but related issues, namely the nature of African environmental philosophy and environmental ethics; the distributive patterns of environmental justice in the North and South; issues of human rights and environmental justice; the land ethic and environmental justice; African relational environmental justice; African ecofeminist environmental justice and African intergenerational conceptions of environmental justice. The chapters seek to provide a coherent, plausible and comprehensive view of environmental justice in African philosophy. Generally speaking, most of the perspectives espoused in this book might be essentially useful to constructing environmental justice conceptions not only in African philosophy but also in other non-African traditions. It is my hope that this book will trigger a quest for more robust engagements and conversations around environmental justice issues across multiple social and cultural divides.

References

Aristotle, (2011). *Nicomachean Ethics* (Trans. Robert C. Bartlett and Susan D. Collins). Chicago: Chicago University Press.

Beer, C. T. (2014). Climate Justice, the Global South and Policy Preferences of Kenyan NGOs. *The Global South*. 8 (2): 84–100.

Behrens, K. G. (2014). An African Relational Environmentalism and Moral Considerability. *Environmental Ethics*. 36 (1): 63–82.

Chemhuru, M. (2021). Land Reform and Redistribution as Environmental Justice Frameworks for Post-Colonial Africa. In, Erasmus Masitera (Ed.) *Philosophical Perspectives on Land Reform in Southern Africa*. Cham: Palgrave Macmillan, 225–240.

Collingwood, R. G. (1946). *The Idea of History*. New York: Oxford University Press.

Conradie, E. M. (2019). A (South) African Land Ethic? The Viability of an Ecocentric Approach to Environmental Ethics and Philosophy. In, Munamato Chemhuru (Ed.) *African Environmental Ethics: A Critical Reader*. Cham: Springer, 127–139.

Cuomo, C. (2002). On Ecofeminist Philosophy. *Ethics and the Environment*. 7 (2): 1–11.

Gyekye, K. (1992). Person and Community in Akan Thought. In, Kwasi Wiredu and Kwame Gyekye (Eds.) *Person and Community*. Washington D.C: The Council for Research in values and Philosophy, 101–122.

Gyekye, K. (2007). *Tradition and Modernity: Philosophical Reflections on the African Experience*. New York: Oxford University Press.

Gyekye, K. (2010). Person and Community in African Thought. In, Kwasi Wiredu and Kwame Gyekye (Eds.) *Person and Community*. Washington D.C: The Council for Research in Values and Philosophy, 101–102.

Gyekwe, K. (2013). *Philosophy, Culture and Vision: African Perspectives*. Accra: Sub-Saharan Publishers.

Kalumba, K. M. (2020). A Defence of Kwame Gyekye`s Moderate Communitarianism. *Philosophical Papers*. 49 (1): 137–159.

Kelbessa, W. (2009). Africa, Sub-Saharan. In, J. Baird Callicott and Robert Frodeman (Eds.) *Encyclopedia of Environmental Ethics and Philosophy*. New York: Gale Cengage Learning, 10–18.

Masitera, E. (2021). Thinking About Land Reform in Southern Africa: The Introduction. In, Erasmus Masitera (Ed.) *Philosophical Perspectives on Land Reform in Southern Africa*. Cham: Palgrave Macmillan, 1–15.

Mbiti, J. S. (1969). *African Religions and Philosophy*. London: Heinemann.

Menkiti, I. A. (1984). Person and Community in African Traditional Thought. In, Richard Wright (Ed.) *African Philosophy: An Introduction*. Lanham: University Press of America, 171–181.

Menkiti, I. A. (2004). On the Normative Conception of a Person. In, Kwasi Wiredu (Ed.) *A Companion to African Philosophy*. Malden: Blackwell Publishers, 324–331.

Metz, T. (2011). Ubuntu as a Moral Theory and Human Rights in South Africa. *African Human Rights Law Review*. 11: 532–559.

Metz, T. (2019). An African Theory of Moral Status: A relational Alternative to Individualism and Holism. In, Munamato Chemhuru (Ed.). *African Environmental*

Moyo, S. (2005). The Land Question and Peasantry in Southern Africa. *Concejo Latinoamericano de Clencias Sociales (CLASCO) Conference Preceding*. Downloaded from: http://biblioteca.clacso.edu.ar/clacso/sur-sur/20100711022553/13_Moyo.pdf. Date Retrieved: 4 August 2021.

Murove, M. F. (2004). An African Commitment to Ecological Conservation: The Shona Concepts of *Ukama* and *Ubuntu*. *Mankind Quarterly*. XLV (2): 195–215.

Murove, M. F. (2020). *African Politics and Ethics: Exploring New Dimensions*. Cham: Palgrave Macmillan.

Mweshi, J. (2019). The African Emphasis on Harmonious Relations: Implications for Environmental Ethics and Justice. In, Munamato Chemhuru (Ed.) *African Environmental Ethics: A Critical Reader*. Cham: Springer, 191–204.

Ndlovu-Gatsheni, S. J. (2018). *Epistemic Freedom in Africa: Deprovincialization and Decolonization*. New York: Routledge.

Obama, B. (2017). https://www.youtube.com/watch?v=30xLg2HHg8Q.

Ramose, M. B. (1999). *African Philosophy Through Ubuntu*. Harare: Mond Books.

Rawls, J. (1971). *A Theory of Justice*. Cambridge: Harvard University Press.

Rolston, H. (1999). Ethics and Environment. In, Emily Baker and Michael Richardson (Eds.) *Applied Ethics*. New York: Simon and Schuster, 407–437.

Samkange, S. and Samkange, T. M. (1980). *Hunhuism or Ubuntuism: A Zimbabwean Indigenous Political Philosophy*. Salisbury: Graham Publishing.

Ssebunya, M., Morgan, S. N. and Okyere-Manu, B. D. (2019). Environmental Justice: Towards and African Perspective. In, Munamato Chemhuru (Ed.) *African Environmental Ethics: A Critical Reader*. Cham: Springer, 175–189.

Warren, K. J. (2000). *Ecofeminist Philosophy: A Western Perspective on What It Is and Why It Matters*. Lanham: Rawman and Littlefield Publishers.

1   Environmental Ethics in African Philosophy

1.1 Introduction

The quest to establish African environmental ethics as an independent discipline in philosophy has in recent times been one of the core objectives in contemporary African philosophy. Only recently has African environmental ethics developed into an independent body, having gained currency in the last half of the twentieth century (Oruka, 1997: xvi). The history of African environmental ethics has otherwise been marked primarily by what was happening to the history of African philosophy prior to the twentieth century, when its existence was denied. This effectively meant that African environmental ethics was also denied, since it is an important aspect of African philosophy. I will not, however, venture into these historical issues of African philosophy in great detail, since that is not my mission here. Nevertheless, I will be constantly referring to it because its history influences that of African environmental ethics. Against this background, there are still some misconceptions as to whether or not African environmental ethics actually exists and whether it is a new discourse altogether. This chapter partly addresses such misconceptions. In that regard, it is important to note that denying the existence of African environmental ethics is as good as denying the existence of African philosophy, since African environmental ethics is an essential part of this philosophy. At the same time, it is absurd to think that way and to even assume or entertain the view that among African communities, there are no conceptions of, for example, the moral status of nature, animal rights and welfare, and the effects of global warming, pollution, climate change, poverty, pollution, the extinction of biodiversity and environmental justice conceptions.

In this chapter, I will dispel some of the views that insinuate that environmental ethics is a new discourse both in the Western tradition and in African philosophy in particular. I seek to show that human beings, particularly African communitarian societies, have always been conscious of the need to relate well with nature, even well before the publication of mid-twentieth century literature on environmental ethics. Such literature – especially that of Aldo Leopold (1949), Rachel Carson (1963), Lynn White (1967), Holmes Roston (1995) and J. Baird Callicott (2002) – suggest that environmental

DOI: 10.4324/9781003176718-2

ethics is a post-modern philosophical discourse. Although these philosophers are not making specific reference to African environmental ethics, they seem to be speaking universally about environmental ethics, implying that even African environmental ethics did not exist prior to the twentieth century. I provide some objections to such views. In order to dispel such misconceptions within the African philosophical context, I trace the history of ancient African environmental philosophy or indigenous African environmental philosophy in terms of how it can be interpreted as conceptualising peaceful existence between human beings and nature. I show how conceptions about various aspects of environmental ethics such as moral status of nature, animal rights and welfare could be read from ancient African philosophy. In order to show the nature of such environmental ethics in African philosophy, I will work with the twentieth century writings on African environmental ethics as well as the twenty-first century discourse on African environmental ethics. The reason for this is that much of ancient African philosophy has not been documented, although some historians of philosophy may want to claim that some aspects of ancient Greek philosophy are actually drawn from African philosophy (James, 1954). I seek to show how this important body of literature on African environmental ethics has largely been more focused on affirming the existence and nature of African environmental ethics against the background of the denial of the existence of African philosophy. I then proceed to consider the missing link in these considerations of African environmental ethics where questions of environmental justice are concerned.

In this first section, I define environmental philosophy and environmental ethics, respectively. I do this in order to be clear about their meanings since I will be making reference to them very often when I examine African environmental philosophy and African environmental ethics in the ensuing sections. In the second section, I examine ancient environmental ethics by situating it within the African philosophical discourse. I intend to counter particular misconceptions relating to the status of both African philosophy and African environmental ethics by analysing the philosophical outlook of African environmental ethics from the ancient African philosophical tradition, the twentieth century and the twenty-first century, respectively. I will end the chapter by demonstrating the research gap relating to the question of environmental justice, which still remains underexplored in these works on African environmental ethics.

1.2 Environmental Philosophy and Environmental Ethics

A clear understanding of both 'environmental philosophy' and 'environmental ethics' is important to this work. I therefore begin by defining environmental philosophy and environmental ethics, respectively. Since I will focus on African environmental philosophy and African environmental ethics in the ensuing chapters, it would be appropriate to first clearly define what environmental philosophy and environmental ethics are before discussing them within the African context. This will be particularly useful when I refer to African environmental philosophy and African environmental ethics, respectively, in questions relating to environmental justice in African environmental philosophy.

Most environmental theorists often use the terms 'environmental philosophy' and 'environmental ethics' interchangeably. However, it must be emphasised that there is a difference between the two, although it remains superfluous and rather less important in much of the discourse on environmental thinking. This is because environmental philosophy and environmental ethics are closely connected concepts by virtue of the fact that environmental ethics is an important domain of environmental philosophy. In my conception of environmental philosophy and environmental ethics in this work, I see no reason to understand and use them separately because they complement each other and are focused on conceptualising good relations between human beings and nature. I therefore seek to show why one need not be distracted by definitional hazy in examining African environmental ethics in African environmental philosophy. Ultimately, my intention is to work on, and interpret African environmental ethics as an important component and complementary area of African environmental philosophy.

Both environmental philosophy and environmental ethics deal with fundamental questions about how human beings ought to relate with nature. Although the central questions of environmental philosophy could be viewed as theoretically broader than those of environmental ethics, just as the questions of philosophy are broader than those of ethics, both at least attempt to search for relations that ought to subsist between human beings and the environment. Robin Attfield defines environmental philosophy as "the study of the concepts and principles relating to human interactions with nature and the natural environment, to related presumptions about the relation of humanity with nature, and to practical implications for both humans and societies" (Attfield, 2018: 38). Although this definition is broader than that with which environmental ethics is actually concerned, it is not too far away, and different from that of environmental ethics per se. Closer to this view of environmental philosophy is also the understanding of environmental ethics as an important branch of environmental philosophy, which is concerned with studying normative values and principles in environmental philosophy.

Environmental ethics seeks to challenge some of the longstanding normative ethical positions relating to human relations with nature. This view of environmental ethics is aptly captured in Aldo Leopold's (1949) notion of the 'Land Ethic', which could be taken as the new theoretical basis for understanding what environmental ethics is all about. According to Leopold, "all ethics so far evolved rest upon a single premise; that the individual is a member of a community of interdependent parts... The land ethic simply enlarges the boundaries of the community to include soils, waters, plants, and animals, or collectively: the land" (Leopold, 1949: 203–4). Understood this way, environmental ethics therefore ought to involve the extension of ethical considerability to the rest of nature or the environment as a whole. Although it is not explicit from Leopold's view, environmental ethics should also consider the possibility of treating the environment as an *end* in itself, or that it should be granted ethical standing for its own sake. In order to take all these fundamental questions into consideration, environmental ethics ultimately ventures into the province of environmental philosophy, such that a clear-cut division between the two might be difficult to make.

The relationship between environmental philosophy and environmental ethics is similar to that of philosophy and ethics, or philosophy and epistemology, where ethics and epistemology are central branches of philosophy. Without either metaphysics, epistemology, logic, aesthetics or ethics, philosophical questions cease to exist. Moreover, philosophical questions can only be philosophical if – and only if – they are either metaphysical, epistemological, aesthetical, logical or ethical in nature. Similarly, all questions of environmental ethics are therefore environmental philosophical questions. However, environmental ethics may not be exhaustive of all environmental philosophical questions, some of which may not really be ethical in nature, although they could have a bearing on conceptions of environmental ethics. By way of example, the metaphysical questions as to whether non-animate beings such as vegetation and physical nature ought to be treated as having ethical status are environmental philosophical questions that they have a bearing, however, on the ultimate conception of environmental ethics. Accordingly, environmental ethics becomes an applied ethical view of environmental philosophy. As an important branch of environmental philosophy, environmental ethics should be understood as practical ethics since it attempts to construct practical conceptions about how human beings ought to relate with the environment. To confirm this practical orientation of environmental ethics, Attfield notes that "environmental ethics studies the principles of values and obligation, the concepts involved, the status of these principles, and their application to practical issues such as the preservation of biodiversity, ecological restoration and the mitigation of climate change" (Attfield, 2018: 39). Given this practical dimension of environmental ethics, I will be making reference to environmental ethics more often than environmental philosophy on most of the issues that I discuss in African environmental philosophy.

Despite the temptation to use the terms environmental philosophy and environmental ethics interchangeably, as I will sometimes be doing here, it must be emphasised that the area of environmental philosophy is broader than environmental ethics in both scope and focus. The area of environmental philosophy is relatively broader in that it considers a broad range of philosophical issues that extend beyond the ethical questions that environmental ethics is specifically concerned with. Environmental philosophy goes beyond the scope of environmental ethics by dealing with philosophical issues that even go beyond merely ethical ones, including questions that might be metaphysical, epistemological, religious, social, political and economic in nature among others (Attfield, 2018: 39). Accordingly, while environmental ethics is concerned with ethical questions relating to value, obligations, principles, application, restoration and applying ethical principles to given ethical scenarios and situations, environmental philosophy goes beyond such ethical questions. According to Katz, "environmental philosophy examines, analyses and (in part) justifies direct ethical principles regarding human action on the natural environment. Its focus is the proper understanding of the relationship between humanity and nonhuman natural world" (Katz, 1991: 80). In short, environmental philosophy and environmental ethics cannot be understood separately.

So far, the generally loose usage of the terms environmental philosophy and environmental ethics to refer to similar questions seems to be clear and somewhat justified. However, in trying to make a distinction between the two, it must be borne in mind that "environmental concerns regularly strain the domain of ethics" (Callicott and Frodeman, 2009: XX). In other words, environmental philosophical questions and concerns are not only ethical but they are quite complex in nature. Ultimately, then, environmental philosophy deals with those questions that fall outside the boundaries of environmental ethics. From a very strict philosophical perspective, for example, one would not imagine that there could be any ethical relations between human beings and nature such that humans could be deemed as having duties and obligations towards it. Such a relationship simply does not and cannot exist. This is because nature (either animate or non-animate) lacks agency in so far as it is not a moral agent in its own right and that it is not our moral counterpart, and therefore lacks reciprocity. Nature could, however, possibly be viewed as some kind of moral patient warranting our attention and care. Moral patients are generally things that are not moral agents and not morally accountable because they lack the prerequisites for such accountability, like all aspects on nature towards which human beings ought to have moral responsibility towards because they can be harmed/made worse off or better by human action (see, for example, Regan, 1983: 19). For this reason, environmental ethics cannot, within its domain properly account for the ethical basis for granting moral status to, for example, non-animate beings and the rest of nature. It might be somewhat better, therefore, to speak of environmental philosophy, which is at least broader that environmental ethics. To bring this view into perspective, Callicott and Frodeman opine that "the silliness of the question *do rocks have moral rights* marks the limits of environmental ethics and poses the need for environmental philosophy to go beyond ethics into the terra incognita of environmental metaphysics, epistemology, aesthetics, and other domains of philosophical inquiry" (Callicott and Frodeman, 2009: xx).

From these perspectives on environmental philosophy and environmental ethics, respectively, there is no doubt that the two concepts are so closely related that it would be difficult to talk of one without making reference to the other. Technically, and strictly speaking, they are different in as far as environmental ethics is a discipline of environmental philosophy. There is a slight difference in terms of what implications each of these disciplines has to environmental thinking, by which I mean the general conception of how human beings, as moral agents, ought to think and relate with the environment. For this reason, I will sometimes loosely use the two interchangeably with reference to African perspectives on environmental thinking. However, more often than not, I will make use of the term 'environmental ethics' even when I am referring to environmental philosophy, as environmental ethics is the term that specifically captures what ought to be the nature of ethical relationships between human beings and various aspects of nature, as well as dealing with the pragmatic dimensions of such a relationship. This explains why the histories of environmental philosophy and environmental ethics are closely connected, as are those of African environmental philosophy and African environmental ethics, as I will demonstrate in the ensuing sections.

1.3 Environmental Ethics in the History of Philosophy

Both environmental philosophy and environmental ethics have a long history, which is also strongly connected to the history of philosophy. Looking at the history of philosophy from the classical or the ancient philosophical tradition, one can see various attempts, from environmental philosophical and ethical perspectives, to conceptualise the nature of the relations that ought to exist between human beings and nature. Katz avers that the literature on environmental philosophy and ethics has a long history stretching as far back as pre-Socratic philosophical times among Milesian cosmologists (Katz, 1991: 79). According to Katz, "...virtually all pre-Socratic philosophers examined the role of humanity in the cosmos, the natural order of the universe. It is no exaggeration to trace the origin of environmental philosophy to the very sources of Western philosophy" (Katz, 1991: 79). Although this view is biased towards Western philosophy, it is wrong to think that environmental ethics is a new discourse that is developing as a result of the current environmental crisis facing the whole world. As I have argued elsewhere, "environmental ethics has traditionally been an inseparable part and parcel of philosophy from the ancient period to contemporary philosophical thinking" (Chemhuru, 2016: 26). Most pre-Socratic speculations by Thales, Anaximander, Anaximenes, Heraclitus, Parmenides and Pythagoras, for example, were attempts to conceptualise the ultimate nature and structure of the universe. For this reason, their speculative philosophical theories on the nature of reality or the surrounding world could be understood, somewhat, as attempts to bridge the gap between humanity and nature and to at least understand nature better. However, some objections may still be raised against this perspective, as some opine that in pre-Socratic thinking, views on environmental ethics are not only lacking in coherence nut also largely anthropocentric. Nevertheless, I will not pursue this argument further than this because it does not fall within the purview of this work, which focuses on the actual African conceptions of environmental justice.

Although the above view that the history of environmental ethics could be traced to as far back as the classical or pre-Socratic period in philosophy is essentially true, my thesis is that the historical origins of environmental philosophy and ethics should not be limited to documented Western philosophy alone. Some non-Western philosophical traditions such as ancient African philosophy also have a lot to contribute to the history of environmental philosophy and ethics through their indigenous philosophies of communitarian philosophy and *unhu/ubuntu* ethics and relational ethics. Western philosophy, however, has tended to be hostile to non-Western philosophical traditions like African philosophy, as confirmed by Hegelian and Kantian thinking. Hegel, for example, perceived Africa as *"Africa proper* and that it has largely remained – for all purposes of connection with the rest of the world – shut up; it is the Gold-land compressed within itself – the land of childhood, which lying beyond the day of self-conscious history, enveloped in the dark mantle of night" (Hegel, 1837/2001: 109). In a similar vein, Kant, in his race theory, opines that Africans, particularly blacks, do not have the capacity for reason. According to Naomi Zack, "Kant was very outspoken about his characterisation of blacks as intellectually inferior, as in another often-quoted remark: *this fellow was quite black from head to foot, a clear proof that what he said was stupid"* (Kant, cited in Zack, 2018: 15). These are both unjustified claims about other philosophical traditions such as African philosophy. However, I will not pay much attention to them because such debates no longer exist in the present day, and because doubting the existence of African philosophy would be self-contradictory. I demonstrate that every cultural tradition has its own intellectual tradition and, ultimately, its own ethic of environmental responsibility. Similarly, classical philosophical traditions in both Western and non-African contexts should, of necessity have had conceptions of how to relate well with nature. Attfield also takes the same view that "some of the relevant concepts, principles, issues and values were used and/or debated during the ancient period, medieval and early modern periods, and proposals for a new approach were made during the mid-twentieth century" (Attfield, 2018: 38), although he still insists that "the conscious and concerted study of these concepts, principles and values first emerged in the 1970s" (2018: 39). For Attfield and other Western environmental philosophers, then, the view that environmental ethics is fairly new is precipitated by quite a number of factors relating to environmental degradation especially after the industrial revolutions. However, I seek to depart from such thinking by locating environmental ethical thinking within the broad history of philosophy, rather than looking at it as something new.

Several accounts on environmental ethics, mostly within the Western philosophical tradition, point to the emergence of environmental ethics in the late twentieth century, especially in the 1970s after the publication of

Rachel Carson's (1963) *Silent Springs*. This classic is viewed as having launched the environmental movement, and consists of essays that drew the world's attention to the looming environmental crisis that posed a threat to both human and environmental wellbeing. In 1967, Lynn White (Jr) also published one of the most influential and thought provoking works on environmental philosophy and ethics, *The Historical Roots of Our Ecological Crisis*. In it, he blames the main strands of the Judeo-Christian heritage as being largely responsible for the overexploitation of nature by human beings, triggering an ecological crisis (1967: 1205). Owing to the influence of these philosophers, much of twentieth century and contemporary discourse on environmental ethics consider these works as marking the beginning of environmental ethics, as expressed by Attfield above. Most of these twentieth century and contemporary environmental philosophers suggest that environmental ethics is something new to humanity. Writing in 1995, Holmes Rolston, for example, opines that "the next millennium is, some say, the epoch of the end of nature. But another hope is that we can launch a millennium of culture in harmony with nature" (Rolston, 1995: 349). Implicit in Rolston's view is the idea that human beings have traditionally not been in harmony with nature and that they can actually *launch* environmental ethics as a new culture.

In Western philosophy, there seem to be many factors influencing the view that environmental ethics is a recent development or a new culture. Some of these factors are related to philosophical questions prior to the industrial revolution most of whose focus were largely speculative, metaphysical and theocentric within the pre-Socratic, Platonic-Aristotelean era and during the medieval period respectively. After the medieval period, more emphasis was placed on the sovereignty of human reason, thereby culminating in the Baconian idea that the human being is at the centre of the universe. This thinking ushered in by the industrial revolution towards the end of the eighteenth century when the detrimental effects on the environment began to be felt on a large scale. As a result, the focus of philosophy shifted to ethical questions again, especially since the effects of the industrial revolution could be felt on a large scale through industrial effluent flowing into the rivers and seas threatening human beings, aquatic creatures and other animals, air pollution from industries and land degradation. However, this should not be understood to mean that prior to these environmental problems, human beings did not have a conception of environmental ethics and justice.

For the most part, the views pointing to environmental ethics as new are also "a result of the growing environmental consciousness and social movements of the 1960s, public interest increased in questions about humans' moral relationships with the rest of the natural world" (McShane, 2009: 407). Although environmental philosophy and ethics could be traced within the history of philosophy itself, these are some of the perspectives that consider it to be a fairly a new discourse. This misunderstanding of environmental philosophy and environmental ethics as new discourses in the history of philosophy is plain to see in the bulk of the literature on African environmental philosophy and ethics as well. This is attributable to the current environmental crisis reflected in climate change, the extinction of species, global warming pollution and heat waves, among other environmental problems. Eric Katz confirms this view as he argues that "it is the recent focus on ecological and environmental problems – the awareness of the ecological crisis in all its manifestations by the intellectual community, government policymakers, leaders of industry and the general public – that has served as an impetus for a specialised literature in environmental philosophy and ethics" (Katz, 1991: 79).

In this section, I have shown how environmental philosophy and ethics are conceived in history of philosophy broadly. Although most of the conceptions considered are largely influenced by Western philosophy, I have cautioned against attempts to consider environmental philosophy and ethics as new. In the next section, I focus on environmental ethics and African philosophy, locating African environmental ethical thinking within African philosophy. Here, I will examine how environmental ethics has been developing within the African philosophical tradition, which has generally been neglected in several accounts on environmental philosophy and ethics.

1.4 African Philosophy and African Environmental Ethics

African philosophy and African environmental ethics have both been denied and excluded from philosophy and environmental ethics. As I have observed elsewhere, "African ethics, and ultimately African environmental ethics, have been determined by the fate and direction of African philosophy in Africa especially after the second half of the twentieth century" (Chemhuru, 2019: 1). Indeed, Masolo shares the same view as he opines that "much of what we have done in the contemporary history of philosophy appears to be only corrective work – that is, to respond to bad philosophy that came out of equally bad scholarship on Africa by European social scientists" (Masolo, 2018: 54), Thus, in considering their historical development, I would not want to also mistakenly assume that both African philosophy and African environmental ethics are new discourses simply because they have been recently included in academic writing. What I seek to demonstrate is how they are becoming more visible recently, and to consider the reasons why they have traditionally been excluded from mainstream discussions of philosophy and environmental ethics, respectively. In order to do this, I briefly consider some historical developments in African philosophy that have shaped the course of African environmental ethics.

Although I do not wish to go deep into the history of African philosophy as such, I find it necessary to at least provide a brief historical background to it. This would enable one to situate environmental ethics within the African philosophical tradition and to locate and trace the conceptions of African environmental justice that I will consider later on in Chapters 2, 3, 4, 5 and 6. As earlier indicated, every philosophical tradition has its perspective/s to environmental ethics, and in this section, I try to show how environmental ethical perspectives in African philosophy have been developing from classical to contemporary African philosophy. To this end, I begin by fleshing out the history, nature and status of African philosophy. I then situate environmental ethics within the history of African philosophy as I consider why African environmental ethics has received scanty attention prior to the twentieth century.

1.4.1 A Look at African Philosophy

African philosophy was generally eclipsed and totally excluded from the philosophical canon until towards the end of the twentieth century. This was largely because the African was seen as a savage incapable of producing any reason or philosophy (Oruka, 1987: 46). Some of the earliest titles produced in the early twentieth century by some colonial missionaries and anthropologists easily confirm this Eurocentric arrogance and exclusivist attitude towards African philosophy. By way of example, Raoul Allier (1929) *The Mind of the Savage* and William Vernon Brelsford's (1935) *Primitive Philosophy*, and William Vernon Brelsford (1938) *The Philosophy of the Savage* all see the African as some kind of a savage only associated with mysterious associations, witchcraft and animistic tendencies. According to Henry Odera Oruka, "these writers found no reason or philosophy in any meaningful sense in what they saw as the 'Mind of the Savage'. Instead, they detected in it a tendency for traditional unanimity typical of the animal instinct" (Oruka, 1987: 46). Such conceptions of Africans have effectively meant that it is not possible to produce African philosophy because in their view, Africans could not by nature be philosophical because they were primitive savages.

Such writings from the early twentieth century dominated philosophical thinking until around the mid twentieth century. It is only after the publication by a Belgian missionary, Father Placide Tempels' (1959) ethnographical work, *Bantu Philosophy,* that African philosophy began to attract attention, at least from non-Africans, who were initially sceptic about it. Tempels can therefore be partly be credited for paving the way for some consideration, albeit ethnographical and colonially oriented interpretation of African philosophical ideas such as African metaphysics, ontology, teleology and ethics within the spaces in which it had been previously denied. However, Tempels' tone and use of terms such as *our Bantu* and *our Africans* (Tempels, 1959: 17 and 24) throughout his work also reflects a somewhat Eurocentric and colonialist mentality characterising his *missionary and civilising* work in Africa. It is quite strange that Tempels and other missionaries in Africa thought that their mission was to introduce religion and thereby civilise Africans. This, despite the fact that African people have always been civilised and religious people notwithstanding the diversity of African religions. Ironically, the purported 'civilisation' of Africa actually meant colonisation. Nevertheless, prior to the publication of Tempels' work, a lot of scepticism surrounded the existence and nature of African philosophy. In addition, even after Tempels' work, much of the debates on African philosophy have been centred on trying to affirm the existence and nature of African philosophy and whether such a philosophy could be *universally tested* and *accepted* into the discourse of *philosophy proper*. However, these pessimisms based on Western universalist fallacies should not be understood to mean that African philosophy did not exist as a philosophy germane and particular to Africa. Nor should it be accepted, that African philosophy could have such a very short history dating back to the twentieth century.

One of the most serious challenges faced in trying to think about African environmental philosophy and ethics is the problem of the universality of knowledge or philosophy. African philosophy has mainly faced the problem of being viewed from within the position of Western universalist arrogance, where it is assumed that Western philosophy must be universally accepted as *the* philosophy proper, and that all other perspectives ought to fit in within this universal framework. As a result, African philosophy was denied because it was thought to be a specific or somewhat second order philosophy that could not meet the requirements to be part of universal philosophy. For example, because of the influence of Tempels' work, there have been attempts, especially by universalist philosophers such as Paulin Hountondji (1996), to reduce African philosophy to ethno-philosophy and use to justify denying it the status as a proper philosophy. This is because it is believed to lack philosophical aptitude, as it is "thought to be deficient in criticality and analyticity, which are considered hallmarks of good philosophy anywhere" (Agada, 2019: 1). Similarly, African environmental philosophy and ethics naturally implicitly faced the same problem of being construed as ethno-philosophy, yet ethics is also sometimes particular and relative to geopolitical spaces and their different communities.

The denigration of African ethno-philosophy has been founded on the notion that there is some kind of a universal philosophy on which other philosophies can be judged. For Agada, however, "the ethno-philosophical stance, like the broader particularist thesis, protests the dogmatic claim of philosophy having an absolute set of determinative criteria discovered in the Western philosophical tradition especially the Anglo-American analytic tradition" (Agada, 2019: 4). Secondly, it must be emphasised that there is no problem with a philosophy being particular. In fact, having as many philosophical perspectives as possible, such as Western, African, Asian, Chinese and American philosophy, is not a problem. If need be, these particular philosophical perspectives might also be reconciled in order to have some kind of universal applicability in solving problems that humanity might face. This is in line with the view that knowledge starts from culture-specific perspectives before it can be seen from a universalist perspective (Agada, 2019: 6). Thirdly, it is even acceptable to have other philosophies within African philosophy itself such as Shona philosophy, *unhu/ubuntu* philosophy, Oromo philosophy and Ibgo philosophy because of the diversity and plurality of African people and worldviews. Such an approach to philosophy is in keeping with what could be referred to as the pluriversality approach to knowledge. Instead of appealing to a one sided and one dimensional view of knowledge and solutions to human problems as implied by the universalist view, the pluriversality view acknowledges the existence of relative and diverse approaches to human knowledge and solutions to problems facing humanity. In this regard, Ndlovu-Gatsheni argues that "pluriversality underscores a world governed by relationality and transcendence over the bourgeois values, knowledge, economic logics, and political perspectives masquerading as scientism and rationality imposed on the rest of humanity" (Ndlovu-Gatsheni, 2020: 24). Following this approach, therefore, the nature of African philosophy and ultimately African environmental philosophy ought therefore to be taken as an alternative approach to knowledge that is more attractive than the universalist view of knowledge, which has tended to deny other pluriversal alternatives and philosophies.

If philosophy is understood as the quest for fundamental wisdom and truths from the diverse and pluriversal perspectives of either metaphysics, epistemology, aesthetics, logic and ethics, then it is a self-contradiction to deny either African philosophy, African ethics, African environmental philosophy or African environmental ethics of existence. Just like Western philosophy, African philosophy should have, and actually has always had, its own metaphysics, epistemology, aesthetics, logic and ethics, including African environmental ethics. The fact is that the histories of slavery, racism and colonialism have tended to eclipse African philosophy from the documented history of philosophy in general. It is for this reason that Chimakonam thinks that slavery, racism and colonialism ignited the emergence of African philosophy in the 1920s as "the colonial order eventually led to angry questions and reactions out of which African philosophy emerged" (Chimakonam, 2014). Although there might be some consensus to this view by Chimakonam that African philosophy emerged as a reaction to slavery, racism and colonialism, I differ profoundly with this view because African philosophy could not have emerged against the background of slavery, racism and colonialism. To accept this view would be to confirm that prior to Africa's encounter with the Western world, there was no African philosophy, or civilisation in Africa. African philosophy, as a philosophy of the African people has existed from time immemorial. If African philosophy is meant to refer to the kind of knowledge generation from the African civilisation and its cultural context, then such a philosophy has existed prior to slavery and colonialism or prior to the encounter between Africa and the Western world. The fact is that slavery, racist and colonial cultures have tended to be hostile to the African cultural and intellectual heritage. If African people have always contributed to their own well-being, identity, civilisation and development (as they have always done), then they have always had their own philosophy and environmental ethics.

Perhaps what should be highlighted with regard to the historiography of African philosophy is how it could be divided into two historical epochs: one made up of undocumented African philosophy (classical/ancient African philosophy) and one characterised by documented African philosophy, made up primarily of mainly twentieth century and contemporary African philosophy. This distinction has always made it difficult to clearly determine the possible starting point for understanding African philosophy and environmental ethics. This also explains why Ademola Kazeem Fayemi opines that "the search for historiography in African philosophy includes dealing with the problem of definition, documentation and method" (Fayemi, 2017: 2). However, my view is that the problem is not as serious as Fayemi thinks because in terms of definition and methods, these questions have already been settled because I have already alluded to the view that African philosophy should not be defined and determined by the Western philosophical perspectives that seek to claim universality.

Undocumented African philosophy could refer to the kind of African philosophy or African philosophical worldviews that have not been written down such as the ancient or classical African metaphysical, epistemological, ethical and political views prior to, for example, the colonial conquest of Africa. The colonial culture in Africa has often served to suppress, distort and destroy this philosophical heritage by presenting it as unphilosophical while legitimising its Eurocentric and *universalist* perspective. In classical African philosophy, for example, philosophies such as communitarian and relational existence, *unhu/ubuntu*, philosophic sagacity, taboo wisdom and consensual political decision-making have always been deployed by traditional African communities as fundamental African philosophies but that they have not been put down in writing. These kinds of philosophies have been preserved and handed down orally from generation to generation despite the purported lack of writing or documentation of African philosophy, which the colonial tendencies have cast doubt on. Entrenched in such tendencies is usually the attempt to distort a people's history by creating notions of what Bruce Janz calls 'non-philosophy and peripherality' (Janz, 2018: 10). Yet, Germany philosopher, Heins Kimmerle reminds us that "there are elements of writing in a mainly oral tradition and elements of orality in a manly literate tradition" (Kimmerle, 1995: 43). Interestingly, in one of the most influential historiographical works on African philosophy, *Stolen Legacy*, George G.M. James (1954) attempts to show that what is rather presented as ancient Greek philosophy is actually African philosophy because of the interactions that ancient Greek philosophers such as Pythagoras, Plato, Aristotle and Alexander the Great had with ancient African civilisations in Egypt. By way of example, with reference to Pythagoras, James notes that "it is said that Pythagoras, a native of Samos, travelled frequently to Egypt for the purpose of his education" (1954: 43). Although James cannot not really prove that the ancient Egyptians were Africans per se, he presents a strong argument to the effect that ancient African philosophy (Egyptian philosophy) could have been popularised in ancient Greece as Greek philosophy. He maintains that "the period of Greek philosophy (640-322B.C.) was a period of internal and external wars, and therefore unsuitable for producing philosophers" (James, 1954: 21). I am also tempted to concur with James' argument because most of the philosophers during this period, especially the earliest Milesian thinkers such as Thales, Anaximander and Anaximenes, merely made metaphysical postulations but could not even elaborate on them.

On the other hand, African philosophy could also be conceived as a systematic discipline consisting of written African philosophy after the 1920s. Chimakonam considers this kind of systematic African philosophy as follows: "thus began the history of systematic African philosophy, with the likes of Aimé Césaire, Léopold Sédar Senghor, Kwame Nkrumah, Julius Nyerere, William Abraham, John Mbiti and expatriates such as Placide Tempels, Jahnheinz Jahn and George James, to name a few" (Chimakonam, 2014). Again, although it is generally referred to as such, to characterise this kind of discourse as *systematic* African philosophy would also be an insult to the works of great African men and women prior to these twentieth century writers because, despite being undocumented, their views on African philosophy are also systematic. This is why Odera-Oruka upholds the need to include of sage philosophy or philosophic sagacity in African philosophy, although he has some reservations about ethno-philosophy (Oruka, 1987). I will therefore use the above characterisation of documented African philosophy as *systematic* under protest. In addition, what makes it even more difficult to take some of these works seriously, especially the second group of thinkers, which Chimakonam refers to as expatriates, is that some of their works have been associated with the colonialist project.

Following these views, African philosophy might be situated and understood from the two respective traditions: one consisting of classical African philosophy contributing to undocumented African philosophy, and the other, which I call the post-modern African philosophy, is made up of documented African philosophy. The latter tradition "draw[s] on the advent of formal education and advent of writing in Africa as the historical starting point of African philosophy" (Fayemi, 2017: 2), although this view remains fundamentally problematic. In the next section, I proceed to show how to interpret African environmental ethics from the two respective traditions of African philosophy.

1.4.2 *African Environmental Ethics*

The plight of African philosophy had a considerable impact on the nature of other important disciplines of African philosophy such as African ethics and African environmental ethics. In particular, African environmental ethics cannot be expected to come from elsewhere except from African philosophy itself. However, because of the Eurocentric misconceptions about Africa, many African ethical value systems, religions and political systems have been largely misrepresented. At worse, they have been viewed as less important than the Eurocentric value-systems. As Kelbessa sees it:

> Limited by preconceived notions, European travellers, missionaries, colonial powers, colonial anthropologists, and some Western philosophers portrayed the African people and their beliefs and practices as barbaric or uncivilised, irrational, unscientific and prelogical. They considered Africans as devoid of morality, religion, and political philosophy
> (Kelbessa, 2009: 14).

Like African philosophy, which has faced such denial, African environmental ethics also suffered the same fate. However, the existence of African environmental ethics, just like that of African philosophy cannot be doubted. Indeed, as Kelbessa observes, "African people have had their own religions and principles for as long as any other peoples... [and] African worldviews embody both anthropocentric and non-anthropocentric attitudes towards the environment" (Kelbessa, 2009: 14). The nature of anthropocentric and non-anthropocentric environmental ethics is clearly shown in both classical African environmental ethics and twentieth century and contemporary African environmental ethics, respectively. While classical African environmental ethics is mainly undocumented, it has been complemented by twentieth century and contemporary African environmental ethics. Most of the African philosophers in twentieth century African environmental ethics and contemporary African environmental ethics start by making reference to classical approaches to African environmental ethics, making it difficult to make a distinction between the two except on the basis of whether or not it is documented.

The historiography of African environmental ethics is quite complex. Classical African environmental ethics in particular is mainly affected by lack of documentation. In this regard, Kelbessa notes that "the lack of documentation before modern times and the variety of peoples in Africa hindered an appreciation of African ethical teachings" (Kelbessa, 2009: 14). However, although much of African traditional environmental ethics has not been documented, unlike the other philosophical traditions, elements of African environmental ethics in ancient African philosophy are plain to see from the oral evidence available through African metaphysical/ontological conceptions of existence, communitarianism and relational living with nature through taboos and totems among others. This could explain why environmental problems in such a tradition are not so pronounced. According to Ikuenobe, "the activities that have raised environmental concerns in Africa did not exist prior to colonialism because Africans had conservationist values, practices, and ways of life" (Ikuenobe, 2014: 2). These elements, values and practices are seen in some indigenous African philosophies of existence that have been passed from generation to generation. By way of example, it is easy to draw from some of the environmental ethical import in philosophies and aspects such as African metaphysical/ontological conceptions of existence, the notion of relational existence through taboos and totems, African communitarian existence, ubuntu and other African maxims and proverbs. Most of these were, and still are, focused on promoting harmonious living between and among human beings as they relate with nature. This is why it is interesting to note that both twentieth century and twenty-first century African environmental philosophy mostly resort to, or at least start from an appreciation of environmental philosophy from the classical tradition. Despite the fact that "many of these traditional conservationist values, ways of life, and attitudes were destroyed by the exploitative ethos of European colonialism and modernity" (Ikuenobe, 2014: 1), most twentieth century and contemporary African philosophers have tried to revive this epistemic tradition and see how it might contribute meaningfully to African environmental ethical thinking.

African ontological, religious and communitarian views of existence have played significant environmental ethical roles in African traditions. Although I would not want to conclude, like Ikuenobe, that in Africa, "environmental problems did not exist prior to colonialism…" (2014: 1), I do at least acknowledge the extent to which African ontological conceptions of existence, religious and communitarian view could go towards inculcating sound environmental ethical thinking. According to such views, African traditional environmental ethics flows from the African communities' metaphysical, ontological, religious, communitarian and ethical orientation of existence. Cameroonian philosopher, Godfrey B. Tangwa observes, for example, that "within the African traditional metaphysical outlook, humans tend to be more cosmically humble and therefore not only respectful of other people but also more cautious in their attitudes to plants, animals, and inanimate things, and to the various invisible forces of the world" (Tangwa, 2004: 387). Understood this way, African metaphysics connects human beings with other visible and invisible forces of the world into some sort of a spiritual relationship that challenges human beings to respect all reality, including the various aspects of the environment. Focusing specifically on the notion of ontology in African traditional metaphysics, Ikuenobe notes that:

> Traditional African views of ontology can be understood in terms of their view of cosmology. Reality is seen as a composite, unity and harmony of natural forces. Reality is a holistic community of mutually reinforcing natural life forces consisting of human communities (families, villages, nations, and humanity), spirits, gods, deities, stones, sand, mountains, rivers, plants and animals
>
> (Ikuenobe, 2014: 2).

From Ikuenobe's view, it is clear that African ontological conceptions of existence have environmental ethical import to human beings as they relate with the various aspects of reality.

This metaphysical and ontological framework of African environmental ethics should not, however, be taken to imply that environmental problems did not exist in such communities. In addition, this particular environmental ethical framework might face some objections for its metaphysical premises. Although Gyekye notes that "moral questions ... may, in some sense, be said to be linked to, or engendered by, metaphysical conceptions of the person" (Gyekye, 2010: 101), Thaddeus Metz (2014) has given a reasonable challenge to this approach and critically challenges fellow African philosophers to think seriously about attempts to ground ethics on metaphysical claims. Notwithstanding such objections, the notion that there are ethical connections between human beings and spiritual beings has always been difficult to justify although African philosophers insist on it. Teleological and vitalist views provide significant support for this metaphysical argument for African environmental ethics. It is based on seeing ontological connections of existence with all the other spiritual beings and various aspects of nature. The African teleological view of existence is almost similar to the teleological view espoused by Aristotle (Aristotle, 2011: NE, Book 1: 1059a). However, Aristotle's view is strongly anthropocentric because of its emphasis on the functional view of a person. In contrast, the African teleological view of existence takes almost all the aspects of reality ranging from spiritual beings, human beings and even physical aspects of reality as also having some purpose for existing. Accordingly, ethical considerability could be extended to all these aspects of reality, albeit sometimes in varying degrees based on their ontological status in the hierarchy of existence in which, for example, human beings could be considered as having more purposive lives than say knives or rocks, which do not actually have life. However, a knife's functional purpose could be that it ought to cut well, and a rock could also serve to further that purpose by sharpening the knife (see Chemhuru, 2019: 33–34). This could be taken as the teleological understanding of African environmental ethics, where all reality has purpose and is therefore deserving of ethical considerability.

Similar to the teleological conception of African environmental ethics is the African vitalist understanding of environmental ethics. This view has mainly been popularised by Tempels (1959), and was later taken up by Magesa (1997) and Bujo (2001). According to such a view of existence, reality is not only teleologically oriented for the sole reason that it has purpose, but that all reality has some vitality or vital force. For Tempels, "there is no idea among Bantu of *being* divorced from the idea of *force*. Without the element *force*, *being* cannot be conceived" (Tempels, 1959: 151–2). Implicit in this idea is the notion that all beings (humans and even physical aspects of the environment) have vitality, or that they are actually vital forces in so far as they have some invisible spiritual energy that is considered to be inherent in them all. (Molefe, 2017: 24). Magesa interprets this vitalist view in African ontology and brings it within the realm of African environmental ethics by arguing that these vital forces are spirits or "active beings who are either disincarnate human persons or powers residing in natural phenomena such as trees, rock, or lakes" (Magesa, 1997: 35–6). Following this vitalist view of African environmental ethics, therefore, the basis for respect and ethical consideration with regards to the environment is that human beings respect the vitality of nature or that nature is imbued with spirituality which humans would not want to create bad relations with.

Over and above the teleological and vitalist views, African environmental ethics can also be understood from the perspective of the African communitarian view of existence. Generally, quite a number of African philosophers have so far presented the African communitarian view of existence as forming the African ethical basis. African nationalist thinkers such as Kwame Nkrumah, Léopold Sédar Senghor and Julius Nyerere can be credited for popularising this view in newly established African democracies in the mid twentieth century. In the African philosophical academy, theologian and philosopher John Mbiti (1969), Nigerian philosopher Ifeanyi Menkiti (1984, 2004) and Ghanaian philosopher Kwame Gyekye (2002, 2007, 2010) have been mainly responsible for interpreting the traditional African communitarian outlook in terms of its import to African ethics in general. These thinkers have demonstrated how the African ontological view of the person is communitarian in orientation. By this, they imply to roughly the view that existence is not individualistic or that the individual cannot live a meaningful and fulfilling life without the surrounding community. By way of example, according to Mbiti, the individual person 'I' can only be understood in terms of one's existence by reference to the community 'we' (Mbiti, 1969: 106). In a similar vein, Menkiti also gives another emphatic view of African communitarian existence, averring that the community overrides the individual's autonomy, including the notion of personhood, which the community can confer to the individual or take away from him or her (Menkiti, 1984). For Gyekye, "the communal or communitarian aspects of African socio-ethical thought are reflected in the communitarian features of the social structures of African societies … [and] … these features are not only outstanding, but the defining characteristics of those cultures" (Gyekye, 2010: 102). Gyekye aptly captures and defends these communitarian conceptions in his moderate or restricted view of communitarianism, which has come to be accepted as the most plausible presentation of African communitarianism (Kalumba, 2020).

These African communitarian thinkers may not have been very explicit in their views about environmental ethics in particular. Notwithstanding, they have managed to read the African view of the person in ways that are making it easier for some contemporary African environmental thinkers to interpret African communitarian philosophy from environmental ethical perspectives. Their views have been taken further and situated within environmental ethical thinking in traditional African communities by such contemporary African environmental thinkers, who suggest that communitarian existence could be further understood from a relational perspective that also takes the environment into consideration. For example, the ground-breaking work on African communitarian ethics, *Unhu/Ubuntuism: A Zimbabwean Indigenous Political Philosophy* by a Zimbabwean couple Stanlake Samkange and Tomie Marie Samkange (1980) has mainly influenced a renewed focus and interest in the African relational ethics of *unhu/ubuntu* and its import to various social and political contexts. In 1999, South African philosopher, Mogobe Bernard Ramose (1999) also published another key text *African Philosophy Through Ubuntu*. In this book, Ramose goes further to demonstrate how the philosophy of *ubuntu* has an ecological dimension, especially in the ninth chapter, *Ecology through Ubuntu*. Overall, these two important works have greatly influenced the direction of contemporary African ethics, which is mostly steeped in the notion of relational existence based on the conception of what it means to exist through *ubuntu/unhu*.

Recently, many contemporary African philosophers writing on African environmental ethics are beginning to interpret African communitarianism and *ubuntu* from varied relational perspectives, by actually situating these philosophies within the context of environmental ethics (see, for example, Murove (2004, 2009), Mangena (2013), Horsthemke (2015), Chimakonam (2018), Chemhuru (2019) and Metz (2019)). These writings on African environmental ethics have so far established that communitarian existence through *ubuntu* and relational existence through taboos and totems play a very significant role in African traditional environmental philosophy. In particular, Murove's (2004) *An African Commitment to Ecological Conservation: The Shona Concepts of 'Ukama' and 'Ubuntu'* demonstrate the relational character of *ukama* (relatedness) and *ubuntu* (beingness/humanness) and the import of these aspects to African environmental ethical thinking (Murove, 2009). In *Taboos as Sources of Shona People's Environmental Ethics*, we demonstrate how the Shona people and other communities in Southern Africa have traditionally used, and continue to use, taboos as prohibitions that guard against the unsustainable use of environmental aspects such as plant species, forests, mountains, water sources and non-human animals (Chemhuru and Masaka, 2010: 121). Fainos Mangena makes a follow-up to this argument and further argues how *mitupo* (totems) and *ukama* (relatedness) discern moral status in the traditional African environment (Mangena, 2013: 26). In the book, *Animals and African Ethics*, Horsthemke (2015) goes further than most interpretations of African environmental ethics by bringing in the aspect of African animal ethics. Although Horsthemke interprets the African communitarian and relational views such as *ukama* and totems from animal ethical perspectives, he is very critical about the essentially anthropocentric practices in these philosophies (Horsthemke, 2015: 25). Another important work is the one by Francis Diawuo and Issifu (2018), although they limit the import of taboo and totem wisdom to Ghanaian traditional communities (which are not the only ones that rely on taboo and totem wisdom in Africa). However, they proffer reasons as to why traditional relational environmental ethics of totems and taboos ought to be infused into policy-making processes in Africa. Drawing from African traditional belief systems based on taboos and totems environmental wisdom, they see "the need to integrate modern laws, traditional customs and norms in the natural resources conservation and management for the benefit the generation yet unborn" (Diawuo and Issifu, 2018: 210). All these works cited in this section confirm the nature of African relational environmental ethics in both the traditional or classical African tradition and contemporary African contexts.

In both classical and contemporary African environmental ethics, it is clear that African indigenous communities are largely environmentally conscious about their well-being and that of the surrounding environment. On the other hand, the African traditional communities' basic attitudes towards the natural environment might sometimes be understood to be somewhat anthropocentric (Horsthemke, 2015: 25). However, African environmental ethics is mostly presented as non-anthropocentric because of its concern with other aspects of the natural environment that include animals and non-animate reality. According to Bujo, "Africans are traditionally characterised by a holistic type of thinking and feeling. For them, there is no dichotomy between the sacred and the secular; they regard themselves in close relationship with the entire cosmos. Total realisation of the self is impossible without peaceful co-existence with minerals, plants and animals" (Bujo, 2009: 281). This ecospheric view of existence among traditional African communities is also echoed by Kelbessa, who notes that "the worldview of most indigenous African communities promotes a unitary conception of reality. In that cosmology human beings are seen as distinct but completely embedded part of nature … balance is seen as the ideal relationship between human beings and the natural environment" (Kelbessa, 2009: 10). This goes to show the extent to which African indigenous conceptions of existence might be taken as sources of African environmental ethics.

Generally speaking, African environmental ethics is based on the respect for the interconnected web of existence. This view of existence also has metaphysical, ontological, communitarian and relational dimensions to it. In this view, all beings, whether physical or metaphysical, are connected to each other and these interconnections and relationships have strong environmental ethical implications. The metaphysical, ontological, communitarian and relational understanding of reality is hierarchically arranged from the Supreme Being, ancestors, human beings, animals and down to the natural environment. Although some might want to interpret this purported hierarchy of existence as anthropocentrically arranged or, at worst, metaphysical, one could also argue that this kind of hierarchy of existence goes to show how existence is interconnected as reflected in Kelbessa's view that traditional "African worldviews are neither wholly monotheistic nor completely anthropocentric. They involve environmentally friendly beliefs and laws that have encouraged or enforced limits to exploitation of biological resources" (Kelbessa, 2009: 10). However, an important environmental ethical perspective that remains missing from the bulk of the literature interpreting African environmental ethics in this way is the question of environmental justice.

In other words, it still remains to be seen how conceptions of, for example, environmental justice for present and even future generations could be read from the African metaphysical, ontological, communitarian and relational views of environmental ethics.

1.5 African Environmental Ethics and the Need for Environmental Justice

In the preceding section, I examined the nature of African environmental ethics in both classical and contemporary African thinking. In this section, I seek to show how the area of environmental justice needs to be further investigated in contemporary writings on African environmental ethics. I note that the focus of most African environmental ethical writings has mainly been placed on its various conceptions, with a particular emphasis on the moral status of the environment, i.e., the living non-human beings (animals), living physical beings (plants species) and physical nature (air, water and soil) (Chimakonam, 2018; Chemhuru, 2019). These and other contemporary writings on African environmental ethics have proved to be important, as they also interrogate some of the contemporary environmental problems facing the African environment, which include climate change, global warming, animal rights and welfare, poverty, pollution and extinction of biodiversity. Amongst these important issues in environmental ethics is the often-neglected notion of environmental justice, which deals with the distributive patterns of environmental benefits and burdens resulting from the environment and environmental change as well (see, for example, the Environmental Protection Agency, 2008).

Although the question of environmental justice is central in addressing injustice, inequality, human rights, marginality and poverty (Bullard, 1990: 2), it has not featured prominently in contemporary writings on African environmental ethics. One of the reasons for this is that the environmental justice movement seems to react to new environmental ethical problems and issues affecting both humanity and the environment. According to Ssebunya, Morgan and Okyere-Manu, "the concept of environmental justice is one of the key issues in environmental debates that has gained recognition among environmentalists in recent times, especially in such a time of climate change, global warming and severe environmental degradation" (Ssebunya, Morgan and Okyere-Manu, 2019: 175). In the Western world in particular, it was popularised in America towards the 1980s owing to problems of waste disposal close to poor black communities (Bullard, 1990: 30). This is why this notion is considered as a fairly new movement, or environmental ethical approach in the late twentieth century. However, within the African context, environmental justice is broadly concerned with various unresolved issues from the colonial history of Africa resulting in disparities between Africa and the rest of the world. Such issues relate to the benefits and burdens accruing from the environment itself, access and lack of access to the environment (land), human rights issues (land dispossessions), health issues (waste disposal), pollution, poverty, food security and the needs and interests of future generations.

In Africa, a number of complex issues demand that African environmental philosophers seriously consider environmental justice issues. The problem of environmental justice is quite complex. It needs to be approached from various perspectives by understanding it, for example, from the historical, social, political and economic perspectives. Addressing issues of environmental justice in Africa would demand an understanding of the history of environmental issues in Africa, and the addressing of some inequalities in the distributive patterns and ownership of environmental benefits and their burdens. Kelbessa captures the complexity of the problems around environmental justice issues in the following observation:

> The colonial legacy, the introduction of a money economy with a capitalist mode of production, the global economic system, the expansion of transnational corporations and commercialisation of knowledge, state control of natural resources, and the appropriation of sacred lands by governments favour Western knowledge and civilisation over the indigenous variety. The expansion of HIV/AIDS, poverty, climate change, and injustice have exacerbated the threat to the existence and development of indigenous environmental ethics
>
> (Kelbessa, 2009: 10).

From this view, problems of environmental injustice in Africa can therefore be traced back to the colonial period, when most indigenous African people were dispossessed of the environment (land) through colonial invasion by the white settler colonial governments after the Berlin Conference of 1884–1885 that regulated European colonisation of Africa by actually dividing and sharing the African continent among European countries such as Germany, Belgium, Spain, Italy, Portugal and Britain.

From the above perspective, the question of environmental (in) justice in Africa could therefore be traced as far back as the beginning of colonialism and slavery. For example, in some African countries to this day, the issue of land ownership remains unresolved. This is why in countries like South Africa, there are still debates around whether to craft legislation to ensure the compulsory acquisition of land from the white minority population without compensation. Even in those countries that have already embarked on some land reform programmes such as Zimbabwe, there are still quite a number of unresolved environmental injustices associated with the land issue. The African environment continues to benefit various transnational corporations that are involved in farming, mining and other economic activities in Africa. Notwithstanding this, indigenous African populations continue to bear the worst effects of environmental injustice from such activities, which include climate change, global warming, pollution and other problems associated with poverty and health problems, HIV/AIDS, food security and the needs and interests of future generations.

When all is said and done, issues of environmental justice in Africa do not seem to have been thoroughly examined in contemporary African environmental ethics. If contemporary African environmental ethics is to be comprehensive, it must take into consideration these aspects and at least demonstrate how such ethics could meaningfully contribute to these topical, yet unresolved, issues in environmental ethics in order to solve primary issues of human rights, marginality and poverty facing indigenous communities in Africa.

1.6 Conclusion

In this chapter, I examined how the quest for African environmental ethics can be conceived in African philosophy. I note that the quest for African environmental ethics should be predicated on a proper interpretation of African philosophy. Despite the history of African environmental ethics being somewhat hazy due to various historical circumstances characterising African philosophy, I also explored some perspectives about how to read African environmental ethics from African philosophy. I ended up considering how African environmental ethics ought to proceed forward by venturing into other less explored issues of environmental justice. To this end, the next chapter examines what environmental justice entails and how it can be conceived in Africa in the light of the disparities between the global North and the global South.

References

Agada, A. (2019). The Sense in which Ethno-Philosophy can Remain Relevant in 21st Century African Philosophy. *Phronimon.* 20 (1): 1–20.

Allier, R. (1929). *The Mind of the Savage.* (Trans. Fred Rothwell.) London: G Bell & Sons Ltd.

Aristotle, (2011). *Nicomachean Ethics: Book 1*: (Trans. Robert C. Bartlett and Susan D. Collins. Chicago: Chicago University Press, 1094a–1095b.

Attfield, R. (2018). Environmental Philosophy and Ethics for Sustainable Development. In, Terry Marsden (Ed.) *The SAGE Handbook of Nature.* London: Sage Publications Limited. Vol. 3 (4): 38–58.

Brelsford, W. V. (1935). *Primitive Philosophy.* London: J. Bale, Sons and Danielsson.

Brelsford, W. V. (1938). *The Philosophy of the Savage.* London: J. Bale, Sons and Danielsson.

Bujo, B. (2009). Ecology and Ethical Responsibility from an African Perspective. In, Munyaradzi Felix Murove (Ed.) *African Ethics: An Anthology of Comparative and Applied Ethics.* Scottville: University of KwaZulu-Natal Press, 281–297.

Bujo, B. (2001). *Foundations of an African Ethics: Beyond the Universal Claims of Western Morality.* New York: The Crossroad Publishing Company.

Bullard, R. D. (1990). *Dumping in Dixie: Race, Class and Environmental Quality.* USA: Westview Press.

Callicott, J. B. (2002). The Pragmatic Power and Promise of Theoretical Environmental Ethics: Fording a New Discourse. *Environmental Values*. 2 (2002): 3–25.

Callicott, J. B. and Frodeman, R. (2009). *Encyclopedia of Environmental Ethics and Philosophy*. New York: Gale Cengage Learning.

Carson, R. (1963). *Silent Springs*. New York: Hough Mifflin Company.

Chemhuru, M. and Masaka, D. (2010). Taboos as Sources of Shona People's Environmental Ethics. *Journal of Sustainable Development in Africa*. 12 (7): 121–133.

Chemhuru, M. (2016). *The Import of African Ontology for Environmental Ethics*. D Litt et Phil (Philosophy): University of Johannesburg. Retrieved from: https://ujcontent.uj.ac.za/vital/access/manager/Repository?query=chemhuru&queryType=vitalDismax&sort=ss_dateNormalized+desc%2Csort_ss_title+asc (Accessed 24 August 2020).

Chemhuru, M. (2019). The Moral Status of Nature: An African Understanding. In, Munamato Chemhuru (Ed.) *African Environmental Ethics: A Critical Reader*. Cham: Springer, 29–46.

Chimakonam, J. (2014). History of African Philosophy. In, *The Internet Encyclopaedia of Philosophy*. Retrieved from: https://iep.utm.edu/home/about/R (Accessed 29 August 2020).

Chimakonam, J. O. (Ed.) (2018). *African Philosophy and Environmental Conservation*. London: Routledge.

Diawuo, F. and Issifu, A. K. (2018). Exploring African Traditional Belief Systems (Totems and Taboos) in Natural Resources Conservation and Management in Ghana. In, Jonathan O. Chimakonam (Ed.) *African Philosophy and Environmental Conservation*. London: Routledge, 209–220.

Environmental Protection Agency, (2008). *Environmental Justice*. Downloaded from: https://www.epa.gov/environmentaljustice (Accessed 15 October 2020).

Fayemi, A. K. (2017). African Philosophy in Search of Historiography. *Nokoko Institute of African Studies*. 6 (2017): 297–316.

Gyekye, K. (2002). Person and Community in African Thought. In, P.H Coetzee and A.P.J. Roux (Eds.) *Philosophy from Africa: A Text With Readings*. New York: Oxford University Press, 297–312.

Gyekye, K. (2007). *Tradition and Modernity: Philosophical Reflections on the African Experience*. New York: Oxford University Press.

Gyekye, K. (2010). Person and Community in African Traditional Thought. In, Kwasi Wiredu and Kwame Gyekye (Eds.) *Person and Community: Ghanaian Philosophical Studies, 1*. Washington D.C: The Council for Research in Values and Philosophy, 101–122.

Hegel, G. W. H. (1837/2001). *The Philosophy of History*. Kitchener: Batoche Books.

Horsthemke, K. (2015). *Animals and African Ethics*. New York: Palgrave Macmillan.

Hountondji, P. (1996). *African Philosophy: Myth and Reality*. Bloomington: Indiana University Press.

Ikuenobe, P. A. (2014). Traditional African Environmental Ethics and Colonial Legacy. *International Journal of Philosophy and Theology*. 2 (4): 1–21.

James, G. G. M. (1954). *Stolen Legacy: The Egyptian Origins of Western Philosophy*. Brattleboro: Echo Point Books and Media.

Janz, B. B. (2018). Peripherality and Non-Philosophy in African Philosophy: Womanist Philosophy, Environmental Philosophy and Other Provocations. In, Jonathan Okeke Chimakonam (Ed.) *African Philosophy and Environmental Conservation*. London: Routledge, 2–23.

Kalumba, K. (2020). A Defence of Kwame Gyekye's Moderate Communitarianism. *Philosophical Papers*. 49 (1): 137–158.

Katz, E. (1991). Ethics and Philosophy of the Environment: A Brief Review of the Major Literature. *Environmental History Review*. 15 (2): 79–86.

Kelbessa, W. (2009). Africa, Sub-Saharan. In, J. Baird Callicott and Robert Frodeman (Eds.) *Encyclopedia of Environmental Ethics and Philosophy*. New York: Gale Cengage Learning, 10–18.

Kimmerle, H. (1995). The Philosophical Text in the African Oral Tradition: The Opposition of Oral and Literate and the Politics of Difference. In, Heinz Kimmerle and Franz M. Wimmer (Eds.) *Philosophy and Democracy in Intercultural Perspective (Philosophie Et Démocratie En Perspective Interculturelle)*. Amsterdam: Rodopi B.V, 43–56.

Leopold, A. (1949). *A Sand County Almanac and Sketches Here and There*. Oxford: Oxford University Press.

Magesa, L. (1997). *African Religion: The Moral Traditions of Abundant Life*. New York: Orbis Books.

Mangena, F. (2013). Discerning Moral Status in the African Environment. *Phronimon*. 14 (2): 25–44.

Masolo, D. A. (2018). History of Philosophy as a Problem: Our Case. In, Edwin E. Etieyibo (Ed.) *Method, Substance, and the Future of Africa Philosophy*. Cham: Palgrave Macmillan, 53–69.

Mbiti, J. S. (1969). *African Religions and Philosophy*. London: Heinemann.

McShane, K. (2009). Environmental Ethics: An Overview. *Philosophy Compass*. 4 (3): 407–420.

Menkiti, I. A. (1984). Person and Community in African Traditional Thought. In, Richard Wright (Ed.) *African Philosophy: An Introduction*. Lanham: University Press of America, 171–181.

Menkiti, I. A. (2004). On the Normative Conception of a Person. In, Kwasi Wiredu (Ed.) *A Companion to African Philosophy*. Malden: Blackwell Publishers, 324–331.

Metz, T. (2014). Questioning Attempts to Ground Ethics on Metaphysics. In, Elvis Imafidon and John A. I. Bewaji (Eds.) *Ontologized Ethics: New Essays in African Meta-Ethics*. Lanham: Lexington Books, 189–204.

Metz, T. (2019). An African Theory of Moral Status: A Relational Alternative to Individualism and Holism. In, Munamato Chemhuru (Ed.) *African Environmental Ethics: A Critical Reader*. Cham: Springer, 9–27.

Molefe, M. (2017). African Religious Ethics and the Euthyphro Problem. *Acta Academia*. 49 (1): 22–38.

Murove, M. F. (2004). An African Commitment to Ecological Conservation: The Shona Concepts of *Ukama* and *Ubuntu*. *Mankind Quarterly*. 45 (2): 195–215.

Murove, M. F. (2009). An African Environmental Ethic Based on the Concepts of *Ukama* and *Ubuntu*. In, Munyaradzi F. Murove (Ed.) *African Ethics: An Anthology of Comparative and Applied Ethics*. Scottville: University of KwaZulu-Natal Press, 315–331.

Ndlovu-Gatsheni, S. J. (2020). *Decolonization, Development and Knowledge in Africa: Turning Over a New Leaf*. London: Routledge.

Oruka, H. O. (1987). African Philosophy: A Brief Personal History and Current Debates. In, Guttorm Fløistad (Ed.) *Contemporary Philosophy: A New Survey. Volume 5: African Philosophy*. Dordrecht: Martinus Nijhoff Publishers, 45–77.

Oruka, H. O. (1997). *Practical Philosophy: In Search of an Ethical Minimum*. Nairobi: East African Educational Publishers.

Ramose, M. B. (1999). *African Philosophy Through Ubuntu*. Harare: Mond Books.

Regan, T. (1983). *The Case for Animal Rights*. Berkeley: University of California.

Rolston, H. (1995). Global Environmental Ethics: A Valuable Earth. In, Richard L. Night and Sara F. Bates (Eds.) *A New Century for Natural Resources Management.* Washington, D.C: Island Press, 249–366.

Samkange, S. and Samkange, T. M. (1980). *Hunhuism or Ubuntuism: A Zimbabwean Indigenous Political Philosophy.* Harare: Graham Publishing.

Ssebunya, M., Morgan, S. N. and Okyere-Manu, B. D. (2019). Environmental Justice: Towards an African Perspective. In, Munamato Chemhuru (Ed.) *African Environmental Ethics: A Critical Reader.* Cham: Springer, 175–189.

Tangwa, G. B. (2004). Some African Reflections on Biomedical and Environmental Ethics. In, Kwasi Wiredu (Ed.) *A Companion to African Philosophy.* Oxford: Blackwell Publishing, 387–395.

Tempels, P. (1959). *Bantu Philosophy.* (Trans. Rev. Collin King). USA: HBC Publishing.

White, L. (Jr) (1967). The Historical Roots of Our Ecological Crisis. *Science.* 155 (3767): 1203–1207.

Zack, N. (2018). *Philosophy of Race: An Introduction.* Cham: Palgrave Macmillan.

2 Environmental (in) Justice in Africa
The North – South Challenge

2.1 Introduction

Across all societies, in order for environmental justice to be achieved, the notions of justice and human rights ought to be universally applicable, observed and binding. Indeed, human rights constitute important elements of virtually any society. However, the question of how to equitably share or distribute the benefits and burdens of environmental or climate change remains unresolved in much of contemporary environmental ethical thinking. This ultimately impacts on the fundamental human rights of affected communities. This problem is related to the central question of justice in society since, according to John Rawls; "justice is the first virtue of social institutions, as truth is of systems of thought" (Rawls, 1971: 3). If justice is properly considered to be one of the first virtues of a good society, environmental justice ought to be considered a fundamental human right in any society. In reality, however, "environmental practices and policies affect different groups of people differently, and environmental benefits and burdens are often distributed in ways that are unjust" (Figueroa and Mills, 2001: 426). As a result, environmental justice issues become closely connected to human rights issues, particularly in the global South, where human rights violations associated with environmental injustice are more pronounced than they are in the global North.

It is also a fact that the world is currently not homogeneous in terms of social, political and economic status and level of human development. Indeed, most countries in the global North are more developed than those in the global South in terms of infrastructure, education, healthcare systems and technological advancements, among others. It is not surprising, therefore, that African communities in the considered less developed and poor global South still need to make use of the Earth's natural environment in order to achieve the kind of development that industrialised economies in the global North have attained. By *global South*, I refer to the less developed regions and countries, most of which are located in sub-Saharan Africa, Latin America and some developing countries in Asia. However, my use of the term *global South* is more specifically a reference to the less developed

DOI: 10.4324/9781003176718-3

and developing regions and countries in sub-Saharan Africa. Although Africa remains behind in terms of development, it is difficult to justify any kind of development envisaged in the global South if it has a negative impact on the environment. In the global North in particular, environmental and climate change issues have now become more urgent than ever before. Africa, meanwhile, is more concerned with its development thrust than with environment and climate change issues. This discrepancy challenges humanity to search for a universally acceptable view of distributive environmental justice based on the understanding that environmental burdens transcend national boundaries (Figueroa and Mills, 2001: 427). However reasonable this might be, this view does not take into consideration, existing disparities in terms of development and distributive patterns of environmental justice.

Central to my argument in this chapter is the question of whether sub-Saharan African communities ought to have duties and responsibilities in terms of climate change mitigation and achieving environmental justice when they themselves need to industrialise and develop their underdeveloped economies. Considering this dilemma, Alexander Gajevic Sayegh argues that "the idea that developing nations should not have duties to mitigate climate change is being progressively abandoned. Yet, the question of how to share the burdens of climate change mitigation is still unresolved" (Sayegh, 2018: 344). In this chapter, I venture into this debate with the intention of revisiting the generally accepted *universal* conception of environmental justice that seeks to understand the economies in the global North and those in the global South as if they are currently in the same *original position*. I use the phrase 'original position' here in the sense in which it is used by Rawls to refer to "the appropriate initial *status quo* which ensures that the fundamental agreements reached [in the quest for justice] are fair" (Rawls, 1971: 15). I contend that the communities in the global North and global South are currently not at par in terms of fair shares in the distributive patterns of environmental justice. I therefore consider how environmental justice issues ought to be understood and distributed from a human rights perspective in societies that have had different social, political and economic opportunities.

There has recently been a considerable body of literature on African environmental ethics. (see for example, Tangwa, 2004; Murove, 2004, 2009; Ikuenobe, 2014; Horsthemke, 2015; Chimakonam, 2018; Chemhuru, 2016, 2019). However, the area of environmental politics concerned with the environmental justice and the human rights of marginalised, poor and black communities has not received much attention from environmental ethicists working in the area of environmental ethics (Bullard, 1990: 2). Accordingly, I try to search for a normative human rights framework on which to base a reasonable conception of environmental justice for sub-Saharan Africa while simultaneously considering the need for development within these communities. I address the central question of whether the developing world in much of sub-Saharan Africa should be allowed to make use of the environment, by whatever means necessary, including some of the provisions of the 2015 Paris Agreement, in order to develop themselves to levels similar to those of developed countries. The Paris Agreement refers to the United Nations Convention on Climate Change Mitigation to which all member countries committed themselves in order to keep global temperatures rises to below two degrees Celsius. This question might be in keeping with John Rawls' idea and conception of justice as fairness in the traditional conception of the social contract (Rawls, 1971: 3), where goods or resources ought to be fairly distributed under appropriate conditions that are acceptable to all.

In general terms, the notion of environmental justice is primarily understood to be a twentieth century movement emanating mainly from the global North. As I have demonstrated in Chapter 1, African environmental ethics is thought to have mainly gathered momentum in the twenty-first century, suggesting that prior to that, the concerns for environmental justice were not considered. For Gonzalez, "while the advantages and disadvantages of human rights-based approaches to environmental protection continue to be debated in the scholarly literature, there is a dearth of research regarding the impact on the North-South power relations of the evolving environmental human rights regime" (Gonzalez, 2015: 152). I therefore suggest how duties, obligations and responsibility relating to environmental justice issues ought to be understood from a human rights perspective. I do this by appreciating the binary distinctions between the global North and the global South and their implications for appropriate conceptions of environmental justice in Africa.

In my discussion, I begin by examining how the notions of (in)justice and environmental (in)justice are generally understood before contextualising them within the African historical context. I then proceed to show how environmental justice ought to be understood as an essentially human rights issue. After that I consider environmental rights from an economic rights perspective. Here, I look at the challenges in contemporary clarion calls and approaches to environmental justice by focusing on the dilemma faced by developing countries in the global South in their quest for development on the one hand while keeping their commitment to environmental justice obligations on the other. I end by proposing an alternative normative framework a sound conception of environmental justice that is in keeping with circumstances of the global South. As I consider and reflect, in general terms, on the paradox of environmental justice in the global South by focusing on the African context, I adopt Rawls' view that "a theory however elegant and economical must be rejected or revised if it is untrue; likewise, laws and institutions, no matter how efficient and well-arranged must be reformed or abolished if they are unjust" (Rawls, 1971: 3) to the extent that such laws do not take the rights of others seriously. This is the position that I have in mind as I propose the plausibility of a human rights-based conception of environmental justice in Africa.

2.2 Understanding (in)Justice

The very idea of environmental justice hinges on how social justice and injustice issues are understood in society. However, the question of justice is quite complex and, has over the years, been approached by multiple scholars using different philosophical perspectives ranging from meritocracy, teleology, virtue, utilitarian, desert, retributive, equality and fairness. These different views reflect approaches to justice that "are not only about how individuals should treat one another. They are also about what the law should be, and about how society should be organised" (Sandel, 2009: 6). Accordingly, I consider some of these perspectives in this section as I seek to trace how justice has been conceptualised in the history of philosophy from the classical period to the contemporary period in African philosophy. I do this in order to reflect on, and hopefully arrive at, what ought to be an acceptable conception of justice that could help us to understand environmental justice and injustice in Africa. I contend that in spite of the various noteworthy perspectives to justice within the history of philosophical thinking, sub-Saharan Africa has faced, and continues to face, the consequences of disproportionate environmental injustice associated with its history of slavery and colonialism that are still unresolved, as I have made reference to in the first chapter. This therefore justifies the search for an alternative view of environmental justice in African philosophy, which is my intention in this book.

As I briefly make reference to history of philosophy broadly, I note that within ancient Greek philosophy, both Plato and Aristotle conceived of justice as the essence of the society and that no society would survive unless it was founded on the proper or acceptable principles of justice. In the *Republic*, for example, Plato proffers a view of justice based on distributing goods on the basis of the virtuous traits and rationality of the individual person (Plato, 1997: *Republic*, Book 1, 2, 3 and 4). For Plato, a just society is a just individual writ large whose tripartite virtues of wisdom, courage and appetites work in harmony. Following Plato's view, it follows that environmental justice is evident where there seems to be harmony in the workings of a person's reason, will and appetites as one relates with various aspects of the environment and other individuals in society at large. A conception of environmental justice in society at large is also determined by the virtues of justice within the individual. However, Plato's view has often been criticised for being ambiguous. Although few would really doubt that justice in the individual is actually a virtue for the individual, the transition of such justice or virtue from the individual person to society is not entirely clear. Plato's organicist view of society, which is based on an understanding of the tripartite conception of persons, is also objectionable for propagating the superiority of human reason, viewed as one of the major supports of anthropocentric thinking.

While Plato's conception of justice has also been opposed for being utopian and exclusivist, Aristotle also weighs in and presents an almost related, albeit different understanding of justice based on the understanding of human happiness and teleology in society. Aristotle suggests that justice entails giving people what they deserve in accordance with what is due to them, whether such persons merit such benefits or desert on the basis of "telos, or purpose, of the good being distributed" (Sandel, 2009: 188). According to this view, for Aristotle, justice might therefore be understood to imply and follow distributive patterns involving *treating equals equal, and un-equals unequal* with regards to whatever is being distributed, in keeping with their differences (Aristotle (2011), NE, Book 5: 140–7). This explains why in the *Politics,* he thinks that women, non-Greeks and slaves do not deserve to be treated equally to Greeks. (Aristotle (2001): *Politics,* Book 1.13: 1280a, 10–25). Aristotle's view of (in)justice has mainly been used to justify some of the Western social and political frameworks responsible for environmental injustice in Africa in the form of slavery and colonialism. According to Sandel, for example, "Aristotle claims that in order to determine the just distribution of a good, we have to inquire into the telos, or purpose, of the good being distributed" (Sandel, 2009: 188). In environmental ethics, this view of justice cannot help us to construct a plausible view of environmental justice because environmental justice demands the equitable distribution of environmental benefits and burdens irrespective of their purpose because all beings ought to have purpose for existing. If, for example, it demands that all countries in the world should contribute to the reduction of air pollution by a certain percentage in order to slow the problem of global warming, this means that the reduction of air pollution should have nothing to do with the purpose for which the polluting industries ought to serve, which is mostly profit-making. In contrast, an action ought to be considered irrespective of its consequences or purposes. For that reason, Aristotle's view of justice is also objected for its relativist, elitist and exclusivist approach to justice and environmental justice at large.

Besides the classical views of justice, there have also been some modern political theorists who have sought fairly egalitarian conceptions of justice among people who are naturally equal. John Stuart Mill's conceives justice within his utilitarian view based on the greatest happiness principle and the idea of liberty of individuals as one of the elements of wellbeing (Mill, 1859/2014: 42). Mill proposes a view of justice that could potentially inform distributive patterns of environmental justice as he thinks that justice is when the individual or society at large receives the shares or distributions to which they are entitled. For Mill, "each will receive its proper shares, if each has that which more particularly concerns it. To individuality should belong the part of life in which it is chiefly the individual that is interested; the part which chiefly interests society" (Mill, 1859/2014: 58). However, even if the individual or society were to be guaranteed the kind of liberty that Mill envisages for them, it is not a guarantee that they will eventually get justice in the end. This is why, appealing as Mill's view might be, it does not necessarily guarantee environmental justice in the practical sense. For that reason, the distributive patterns of environmental justice between and among individuals and between the global North and the global South are still not fair.

In *Justice as Fairness*, John Rawls offers an alternative view of justice to the utilitarian perspective, one based on trying to understand all individuals as free and equal in the *original position* (Rawls, 1971: 15). Such individuals will also have basic rights and liberties, and the prioritisation of these basic rights and liberties is the first objective of justice as fairness. A second objective was to integrate fair equality of opportunities with the difference principle in order to achieve fairness and equality. Rawls' view of justice is very useful to the conception of environmental justice. According to Rawls, "justice denies that the loss of freedom for some is made right by a greater good shared by others. It does not allow that sacrifices imposed on a few are outweighed by the larger sum of advantages enjoyed by many" (Rawls, 1971: 3). From Rawls' viewpoint, justice ought to imply equality and fairness in the distributive patterns of goods across all societies. Implicit in the same view is the idea that as long as the individual or the society are suffering from the unequal distribution of environmental goods, benefits and burdens, then, effectively there is injustice and ultimately environmental injustice. Unfortunately, despite the appeal of Rawls's theory the world over, environmental benefits and burdens are unjustly distributed. Rawls's theory of justice is therefore more of a utopian theorisation of social and political institutions. This prompts Amartya Sen to envisage a view of justice complementary to that of Rawls, in a quest for principles of justice based on the lives, capabilities and freedoms of the people involved rather than institutions (Sen, 2009: xii), since institutions do not always guarantee justice. Human capability or capacity could at least be some sort of power that could be used in the search for justice as a precondition for human development (Sen, 2009: 253–268). Understood this way, human capabilities and freedoms become central in the search for environmental justice as well.

Unlike the Platonic and Aristotelean theories of justice, at least these "modern theories of justice try to separate questions of fairness and rights from arguments about honour, virtue, and moral desert. They seek principles that are neutral among ends, and enable people to choose and pursue their ends for themselves." (Sandel, 2009: 197). Accordingly, the theories of Mill, Rawls and Sen offer fresh perspectives on how to distribute justice in society, including environmental justice based on certain rights, claims or entitlements, capabilities and freedoms that human beings should have by virtue of their status as human beings. Although I am concerned with African conceptions of environmental justice, these approaches will also be useful to my approach because of their emphasis on the rights and entitlements that individuals ought to have.

2.3 The Historical Roots of Environmental (in)Justice in Africa

Notwithstanding the diverse approaches to justice cited above, some of which are useful in my conceptualisation of plausible conceptions of environmental justice, sub-Saharan Africa has faced various forms of injustice, human rights violations and environmental injustice. It is worth noting that most of these injustices and rights violations are mainly attributed to the role of the Western world in Africa. Since the sixteenth century when slave trade started in Africa, injustice has always been a characteristic feature of the Western world's treatment of black African people. Some of these injustices are still being be felt today, including environmental injustice, which is affecting sub-Saharan Africa more than it is the global North.

In Africa, injustice was mostly experienced through slavery which saw African people subjected to inhuman treatment and shipped like sardines to work as slaves in plantations and mines in Europe, Asia and America. This marked the beginning of injustice and environmental injustice against black Africans, who were forcibly removed from their land. It was also to be worsened by colonialism, which started in Africa towards the end of the nineteenth century, subjecting sub-Saharan Africans to injustice and fundamental human rights violations as they were dispossessed from their land and properties. As I have argued elsewhere:

> The land question and other problems associated with it in contemporary Africa such as dispossession, land imbalances, poverty and environmental injustice could all be attributed to the Berlin conference of 1884–1885. It is at this conference that, owing to Africa's vastness in terms of size, favourable climatic conditions and natural resources, it became a prime target for European countries such as Germany, Belgium, Spain, Italy, Portugal, and Great Britain
>
> (Chemhuru, 2021).

Colonialism has therefore largely contributed to some of the human rights violations including, the right to life and liberty, which Mill sees as some of the most basic human rights that humans ought to be guaranteed.

After the advent of colonialism in Africa, the indigenous population was dispossessed of their property (land) including their livestock, thus violating their fundamental right to human dignity and property, particularly the land, which is very central to human life and well-being. Although I do not agree with Aristotle's conception of humanity because of his justification of slavery, I do accept his observation that "property is part of the household, … and no man can live well, or indeed live at all unless he be provided with necessaries" (Aristotle, *Politics*, Book V: 156). Because of this dimension, the land issue and property rights have been some of the root causes of liberation struggles against colonialism in Africa. According to Mawondo, "a deep sense of injustice caused by inequalities and deliberate dispossession of Africans by the white settler regimes was amongst the fundamental causes of the liberation struggle in Zimbabwe" (Mawondo, 2008: 7). Although Mawondo is more focused on Zimbabwe, this form of environmental injustice and inequalities created through land dispossessions by the white settler regimes occurred throughout the African continent.

Since communities in sub-Saharan Africa have, for the most part been deprived of the fundamental right to property and ultimately the right to the environment, they have been exposed to poverty. Although poverty is not always the result of injustice (Mawondo, 2008: 9) in Africa, it is primarily attributable to injustice, and in particular, to environmental injustice created by colonial structures. This accounts for the view that poverty in Africa is largely a result of "concrete historical acts of deliberate dispossession ... and hence an injustice" (Mawondo, 2008: 9). This has precipitated the social, political and economic disparities between the North and South, culminating in social and environmental injustice in sub-Saharan Africa, a problem which is very complex and remains unresolved in post-colonial Africa.

As I have shown above, in post-colonial Africa, the problem of social and environmental injustice resulting from land and property dispossessions still persists. At independence, most African governments adopted some reconciliatory approaches with their former colonisers, which did not address injustice caused by land dispossessions. This explains why there is still widespread injustice, poverty as well as equalities in most sub-Saharan African countries. These problems mainly affect indigenous African populations, who still do not have access to land. Although Zimbabwe tried to solve this particular problem at the turn of the new millennium, the process was so haphazard and chaotic that it ended up creating more injustice, poverty and inequalities amongst the indigenous populace, because the land reform programme only benefitted a few politically connected individuals. In other countries such as South Africa, such inequalities are even worse because governments have not yet started to address them.

Within the global South, and particularly in sub-Saharan Africa, the quest for environmental justice ought to be approached by considering the human and economic rights of the less developed communities in this region. This quest for environmental justice in Africa should not be an African affair alone, in which only African people search for plausible ways to achieve environmental justice. The global North ought to have some obligations in addressing the problem of environmental injustice in Africa because environmental problems in Africa are largely attributed to its colonial history and the continued pollution from the extractive industries of transnational corporations in Africa, which have their headquarters in the developed countries of the global North. For these reasons, environmental justice is a fundamental human and economic rights issue in Africa because of the interplay of various factors such as inequality and deprivation (from the environment/land), unequal treatment and the non-involvement of everyone in environmental decision making.

2.4 Environmental (in)Justice in Africa from a Human Rights Perspective

The movements representing the debates on the quest for environmental justice are quite complex. In the United States of America, the term environmental justice became popular when it was used by activists and lobbyists in their protests against waste dumping in North Carolina in 1982 (Bullard, 1990: 30). One of the most famous activist scholar on environmental justice issues, Robert Bullard popularised the movement with the publication of his (1990) book, *Dumping in Dixie: Race, Class and Environmental Quality*. This important work focuses on the social and political issues associated with environmental justice, although it is focused on a different context to one that I am concerned with in this book. However, most of the important observations that it raises, such as the connections between race, class, power, politics and on the other hand, environmental justice, are all important to my conceptualisation of environmental justice in African philosophy. As I consider aspects such as relations between the environment and race, the environment and poverty, the environment and economic development, I seek to conceptualise a plausible human rights-based view of environmental justice for sub-Saharan African communities, most of which are still poor and underdeveloped. I will work within the context of environmental justice as understood from a distributive justice perspective but one that also ought to be informed by a conception of environmental justice as a human right. From this perspective, I intend to work within the context and line of other environmental justice movements and scholars "who have worked to integrate considerations of distributive justice into areas of environmental law..." (Purdy, 2018: 810) and human rights. Indeed, I will show how the distributive patterns of environmental justice might affect human rights conceptions in general, as well as how to uphold human rights through sound environmental policies and practices.

So far, the human rights discourse as articulated in the 1948 Declaration of Universal Human Rights and subsequently the African Charter on Human and People's Rights in 1981 appears to be silent about how to guarantee, safeguard and claim human rights violated as a result of environmental injustice across the world. Indeed, it is a fact that human societies that face environmental injustice also face fundamental human rights violations. Gonzalez submits the view that:

> Although most human rights treaties do not explicitly recognise the right to a healthy environment, global and regional human rights tribunals have determined that inadequate environmental protection may violate the rights to life, health, food, water, property, privacy and the collective rights of indigenous peoples to their ancestral lands and resources
>
> (Gonzalez, 2015: 156).

The best way to understand environmental justice would therefore be first to address the claims for human rights violations, particularly those relating to access to the environment and the natural resources therein. In Africa in particular, the quest for environmental justice should be approached through the demand for human rights that takes into account its colonial past which, according to Gonzalez, "still persists in one form or another to the present day" (Gonzalez, 2015: 160). The way in which colonial past is still persisting in Africa could be witnessed in the dominance of foreign investors in African economies. However, I will not get into detail about this perspective as it does not add value to my discussion here.

The notion of distributing environmental justice must also be understood as a global concern in response to unpersuasive global environmental practices and policies that support the inequitable distribution of environmental benefits and burdens, violating fundamental human rights. Environmental justice is a human rights concern in the sense that problems associated with inequitable distribution and inequalities pertaining to environmental or natural resources are still widespread, affecting different people such as the poor, racial and ethnic minorities, women and children in various ways. Gonzalez argues that "the adverse impacts of the global environmental degradation are borne disproportionately by the planet's most vulnerable human beings, including the rural and urban poor, racial and ethnic minorities, women and indigenous people" (Gonzalez, 2015: 154–5). It is for this reason that environmental issues are essentially human rights issues. Unless these human rights issues are addressed, these vulnerable and most affected will always face human rights violations through environmental injustice.

Ultimately, environmental injustice impacts on people's basic human rights to land and all other natural resources necessary for human livelihood. Similarly, the threat of environmental deterioration and its impact on both humanity and nature itself are being felt by all human beings irrespective of their geographic, social, political and economic position. It is interesting to note, however, that environmental justice issues in the global North are primarily motivated by the reality of environmental deterioration and not necessarily an independent social, political or ethical framework based on human rights as is the case in the global South. Although the notion and discourse of environmental justice are primarily influenced by the threat to the natural environment through pollution, climate change, global warming, etc., the quest for environmental justice also ought to be understood as an important element in the quest for global justice, and ultimately human rights, mostly among communities in the global South. This is because environmental justice is concerned with the whole environment as a basic human right and an entity to which everyone is entitled. For this reason, the natural environment becomes a rallying point for global justice issues, human rights because human beings across the globe interact differently with the natural environment, affecting others in different ways. According to Bill Lawson:

> Environmental justice, at least, entails preserving the environment as a global entity, but also making those persons who feel, have felt, have been, or are victims of environmental crimes and atrocities feel as if they are part of the solution as full members of the human community and not just the environmental dumping ground for the well-off
>
> (Lawson, 2008: 1).

Globalisation and global justice still remain some of the strongly contested notions in the world, although I will reserve their discussion here. Nevertheless, environmental justice issues have strong implications for these important aspects.

Notwithstanding the contestations around the understanding of globalisation, global justice, human rights and environmental justice, I will take one of the universally accepted and articulated views of the notion of environmental justice that is provided by the Environmental Protection Agency in America as follows:

> Environmental Justice is the fair treatment and meaningful involvement of all people regardless of race, color, national origin, culture, education, or income with respect to the development, implementation, and enforcement of environmental laws, regulations, and policies. Fair Treatment means that no group of people, including racial, ethnic, or socioeconomic groups, should bear a disproportionate share of the negative environmental consequences resulting from industrial, municipal, and commercial operations or the execution of federal, state, local, and tribal environmental programs and policies. Meaningful involvement means that: (1) potentially affected community residents have an appropriate opportunity to participate in decisions about a proposed activity that will affect their environment and/or health; (2) the public's contribution can influence the regulatory agency's decision; (3) the concerns of all participants involved will be considered in the decision-making process; and (4) the decision makers seek out and facilitate the involvement of those potentially affected
>
> (Environmental Protection Agency, 2008)

Although this view has generally been taken and accepted as the universal understanding of environmental justice, it is quite difficult to meaningfully conceptualise a smooth and all-encompassing global view of environmental justice in the world because of various factors associated with levels of human development, poverty, race, class and gender, which are essentially human rights issues. This challenge of coming up with an acceptable view of environmental justice also emanates from the very problem faced in trying to come up with universal conceptions about issues such as justice and human rights in the world. However, such an exercise is worthwhile.

One of the major challenges of trying to universalise a conception of environmental justice is that the question of distributing environmental justice is closely connected to issues of racism, sexism, classism, poverty and human rights, all of which cannot be easily resolved. Among Afro-American black communities in America, for example, Robert Bullard observes that "a growing number of African American grassroots activists have challenged public policies and industrial practices that threaten the residential integrity of their neighbourhoods" (Bullard, 1990: XVI). In most African communities, given that issues to do with race, class, gender, poverty and human rights remain unresolved, it is quite difficult to meaningfully conceptualise a view of environmental justice that is comparable to other non-African communities in different circumstances. By way of example, colonial land imbalances primarily based on race, class and gender still exist in most post-colonial African countries. Unless such human rights related questions are addressed first, it might be difficult to construct a plausible view of environmental justice. In the global South, the quest for environmental justice might therefore make sense if it is approached in a manner that upholds human rights.

From the above discussion, therefore, it is clear that as long as our societies continue to accept environmental policies and philosophies that promote racism, classism, poverty injustice and human rights violations, then it will be very difficult to conceptualise and realise a meaningful view of distributive environmental justice. As Figueroa avers, "concerns for the distributive dimension of environmental justice begin with the observation that people of color, the poor, and under-represented groups such as indigenous tribes and nations are faced with a disproportionate amount of environmental burdens" (Figueroa and Mills, 2001: 427). This, is despite the fact that the poor and less developed countries in Africa contribute less towards environmental damage through pollution as compared to their counterparts in heavily industrialised countries in the global North. However, these countries bear more of these environmental burdens and at the same time are expected to conform to global environmental policies. (Kelbessa, 2015: 45). In the next section, I seek to highlight some of the dilemmas that African communities face vis-à-vis the discourse of global environmental justice.

2.5 Economic Rights, Environmental (in) Justice and Africa's Dilemma

World economies are generally ranked within three major categories; i.e., the developed or first word countries (Australia, Canada, France, Germany, Italy, Norway, Sweden), the newly developed countries (Brazil, China, India) and the developing countries (the majority of sub-Saharan African countries, with the possible exception of South Africa, which exhibits the characteristics of both a developed and a developing country). In this work, however, I will bracket all sub-Saharan African countries as belonging to the category of developing countries in the global South, despite notable differences in terms of economic development therein.

Despite the above categorisation of world economies according to both geographical and economic categories, they have in common the fact that they are threatened by environmental degradation and climate change. These problems are triggered by pollution through the emission of various toxic substances and emissions into water bodies and the atmosphere. In the face of such problems, world economies in both the global North and South are uniting to fight this phenomenon from a common front, advocating for environmental justice because "climate action is no longer marked by a sharp binary north–south division" (Sayegh, 2018: 344). Indeed, the effects of climate change are evident in floods, droughts, global warming and emergence of new diseases like COVID-19. However, the most affected populations remain the poor, disadvantaged and minority communities, primarily in Africa and the global South at large (Minority Rights Group International, 2008: 1), which are yet to attain their fair share of economic rights such as land, housing, food, water, health care and education. Indeed, the greatest burden of environmental injustice currently affecting these communities can be attributed to human rights violations of poor and disadvantaged people in the global South through the colonisation of Africa. According to Gonzalez, "the roots of contemporary environmental injustice live in colonialism. The European colonisation of Asia, Africa, and the Americas devastated indigenous societies and wreaked havoc on the flora and fauna of the colonised territories through logging, mining and plantation agriculture" (Gonzalez, 2015: 159). In order to address such injustice, that is still manifest today, the human rights situation of such communities should be taken into account. From a human rights perspective, such communities ought to have what Masitera views as economic rights, which "pertain to the freedoms that individuals should have so as to attain reasonable livelihoods in society" (Masitera, 2018: 19). It would not make sense to seek ways of contributing to environmental justice itself without having basic access to the environment itself.

In addition, it is also striking that those responsible for the highest levels of emissions are located in the developed countries and the newly developed economies in the global North, respectively (Sayegh, 2018: 345). At the same time, developing African countries in the global South have relatively few economic rights and relatively little access to the environmental resources. Their levels of emissions are insignificant because of their lower levels of industrialisation and development. As a result, it is now evident that almost one-fifth of the world's human population consumes almost four-fifths of the world's natural resources, leaving about four-fifths of the world's population with about a fifth of the world's natural resources (Figueroa and Mills, 2001: 426). Ultimately, those who consume more of the world's natural resources also produce more in terms of its industrial effluent. The effect of this disparity is mainly felt by the less developed or developing countries, which are on the receiving end of many of the effects of climate change. As Kelbessa argues in a similar vein, "although climate change affects all countries, the world's poorest countries face the most severe impact. It is the wealthy who are using most of the energy that leads to the emissions that cause climate change, while it is the poor who will bear most of the costs" (Kelbessa, 2015: 45). Despite this disparity in terms of contribution to environmental pollution and climate change, the quest for environmental justice demands that all countries have equal obligations and responsibilities towards the attainment of environmental justice. Yet, African countries still need to consider the aspect of entitlement to economic rights and economic means so that they can at least have "opportunities to realise basic needs necessary to function in a society" (Masitera, 2018: 19).

The disparities in the access to economic benefits, and enjoyment of nature's benefits and burdens, as well as the urgent calls for environmental responsibility naturally raises the question of whether there is fairness in the use of the world's natural resources and the resultant emissions of pollutants into the atmosphere. Although Bullard is more focused on the black American communities in North America, he makes an important observation about the injustice and ultimately the deprivation of human basic rights experienced primarily by the less privileged and poor communities across the world. According to Bullard, "an abundance of documentation shows blacks, lower-income groups, and working-class persons are subjected to a disproportionately large amount of pollution and other environmental stressors in their neighbourhoods as well as in their work places" (Bullard, 1990: 1). This is not only an issue of environmental injustice but it also affects the basic human rights and entitlements of such people to a clean working place and environment.

Ownership of and entitlement to economic means of production such as natural resources constitute a basic human or economic right that also contributes to the lives and well-being of individuals. However, most owners of the means of production in African countries are not the indigenous communities. Rather, "much of the ecological harm in the global South is due to export oriented production rather than domestic consumption and to unsuitable natural resource exploitation by transnational corporations" (Gonzalez, 2015: 154). As a result, mining and industrial activities in the global South communities are mostly located close to marginalised and poor communities, which do not, however, benefit from any of these activities. By way of example, when diamonds were discovered in the Marange area in Manicaland province of Zimbabwe after the turn of the new millennium, local communities in the Marange area were displaced and relocated to the periphery of these mining fields where they continued to endure poverty, displacement, pollution and other environmental stressors from these mining activities.

The other irony of the idea of environmental justice as a question of human rights lies in its prescriptive use by the developed world in the emerging and developing countries, whose communities are, for the most part, still poor.

In most cases, the discourse on the need for environmental justice also seems to be coming mainly from those countries responsible for the highest emission levels whose economies developed long before the emerging or developing countries of the global South. Accordingly, Figueroa and Mills argue that

> the poor and developing nations around the world (primarily in the global South) attempt to satisfy their rights to development as the rich, industrialised nations (primarily in the global North) call for environmental protection against the same development practices that they themselves invented and used for decades and which introduced much of the environmental degradation we see around the world today
>
> (Figueroa and Mills, 2001: 426).

What is paradoxical here is the universal quest for fairness with regards to the use of nature's natural resources and the need for the equal responsibility towards environmental degradation. From this view, both the developing and developed countries ought to have such a responsibility without taking into consideration the disparities in the levels of human and economic development.

I have already emphasised above, within the developing world, and Africa in particular, environmental injustice has actually been committed not by these African countries themselves, but mostly by the majority of those in the developed European countries through colonialism as. Ikuenobe argues, in the same vein that "one motivation for Europeans in the colonisation of Africa was the need to get raw materials for their industries and the need to get markets for the products of their industries" (Ikuenobe, 2014: 12). Colonialism has therefore been largely responsible for environmental degradation and environmental injustice in Africa because it has contributed to environmental disparities between the North and the South through the plunder of natural resources in Africa as well as Africa's traditional value systems that are aimed at safeguarding sound environmental ethical thinking. Ikuenobe intimates, in this regard, that "many of these traditional African values, ways of life, and the moral attitudes of conservation were destroyed by the exploitative ethos of European colonialism and modernity" Ikuenobe, 2014: 2). Accordingly, this scholar concludes that "colonial structures have created for Africa very interesting problems that have a bearing on the current environmental problems in Africa" (Ikuenobe, 2014: 13). It is for this reason that this book, partly, seeks to challenge African philosophers to revisit, and make use of these African philosophies in order to at least confront some of these colonial structures that inhibit the attainment of environmental justice in Africa.

Overall, it must be emphasised that the impact of colonialism on Africa has largely contributed to injustice and economic rights violations that are still ongoing. To confirm this view that colonialism did not end with the attainment of political independence in Africa (Nkrumah, 1965: 197–201), several transnational corporations dominate the exploration of Africa's vast mineral and other natural resources, which they use to industrialise and grow their own economies in the global North. China has recently become a major player in this regard. This explains why, for example, "the most industrialised nations of the world have produced enough CFCs to generate dangerous holes in the ozone layer of the atmosphere, with deleterious effects on the health of the entire world's plant, animal, and human population" (Figueroa and Mills, 2001: 426). Paradoxically, it is these very same industrialised nations that *preach* the *gospel* of environmental justice to societies whose economic rights have been violated.

From the above-mentioned view, it therefore becomes very difficult to speak about, and even envisage a universally accepted view of environmental justice in a world that is not socially, politically and economically homogenous. Attaining environmental justice remains elusive for as long as gender, class, racial, social, political and economic divisions and inequalities that currently characterise the world are not resolved. Lawson avers that "the problem is the bringing together of diverse environmental stakeholders to resolve issues regarding the environment. This is particularly difficult when environmental policies appear to be rooted in class or race divisions" (Lawson, 2008: 2). This view is further compounded by the fact that women, people of colour, the poor and the less developed nations in the global South are the ones that endure, for the most part, the burdens of environmental damage and policies that have benefitted most countries in the global North. This brings in the question of human rights violations that are mainly that are mainly attributable to environmental degradation contrary to the provisions of the various international covenants on civil, political, economic, social and cultural rights including the African Charter on Human and People's Rights (Gonzalez, 2015: 156). One can therefore come to the conclusion that the proclamation of these rights provisions have not been characteristically synonymous with the end of injustice and human rights violations mostly in the global South. Nevertheless, these rights provisions remain important in the safeguarding of human rights and justice, including environmental justice.

2.6 Prospects for Distributing Environmental Justice in the Global South

To this point, an acceptable view of distributive environmental justice seems difficult to arrive at, if one considers the complex binary division between the global North and the global South. Such disparities impede the equitable distribution of environmental benefits and burdens or even the meaningful involvement of all people in the implementation and enforcement of environmental laws and policies. Even if one accepts Lawson's view that "the persons responsible for wrongdoing have the greater responsibility for correcting it" (Lawson, 2008: 3), it is still difficult to arrive at an acceptable view of environmental justice. Lawson's view suggests that the global North has the responsibility for environmental justice in the global South because it has long been responsible for violating the environmental rights of such communities.

However, if environmental justice is understood as the equitable distribution of the benefits and burdens emanating from environmental and climate change as well as the involvement of all people in environmental policy implementation regardless of race, colour, origin, culture and education (Lawson, 2008: 1) in order to solve the current dramatic changes on the planet, there is an urgent need for an acceptable compromise or view of environmental justice. In order to arrive at a plausible view of environmental justice, I partly make reference to John Rawls's two principles of justice that are assumed to be the only choices for all self-interested rational beings behind the veil of ignorance (Rawls, 1971: 52–3). I apply Rawls' theory to the African context so as to construct a normative environmental justice framework that could take into account the rights of the people in the global South.

First, a distributive approach to environmental justice could be the starting point in formulating a plausible environmental justice framework that is acceptable to the global North and African communities. All human beings should strive to be, or revisit Rawls' point of the *original position* (Rawls, 1971: 15) to conceptualise an acceptable view of environmental ethical thinking embracing environmental justice for both the global South and the global North. Otherwise, it is difficult for human beings to come up with a uniform view of environmental justice if they are not in the same original position, and that their veil of ignorance would be a rushed position as is the case with contemporary clarion calls for global environmental justice (Rawls, 1971: 101). My view is that the global quest for environmental justice supposes an original position based on original freedom and equality among human beings and their communities. Yet this is not the case in most African countries, given the history of slavery, colonialism, racism and apartheid. Because of this history and its effect in perpetuating inequalities, injustice and human rights violations in Africa mostly by its counterparts in the global North, environmental justice is difficult for Africa to attain for as long as these issues remain unresolved. Accordingly, Rawls' view of the original position as the starting point for thinking about distributive justice might at least set the tone for a fair distributive justice framework that is be acceptable in Africa.

If environmental justice is to be meaningful to the African condition, certain minimum standards of fairness and equality should be satisfied. Although Rawls's argument is not specifically focused on the African context per se, he does articulate the need to satisfy these conditions observing that:

> The intuitive idea of justice as fairness is to think of the first principles of justice as themselves the object of an original agreement in a suitably defined initial situation. These principles are those which rational persons concerned to advance their interests would accept in this position of equality to settle the basic terms of their association. It must be shown, then, that the two principles of justice are the solution for the problem of choice presented by the original position
>
> (Rawls, 1971: 102).

Based on this perspective, my contention is that although environmental justice is an urgent issue, the envisaged conditions for the original position are not fully satisfied. By way of example, according to Rawls, one of the options for individuals in the original position is that they ought to choose from alternatives, which are open to all persons in the original position. Similarly, although the quest for global environmental justice is urgent, there must be some way in which the questions of injustice and human rights are addressed first for economies in the global South so that everyone is in what Rawls (1971: 102) calls a *suitably defined initial situation*. The point is that human rights and human development are inseparable components of the quest for justice. Therefore, in order to successfully engage, and take on body, the less developed countries in Africa, some effort must be done to develop their economies first so that they reach where the developed countries are presently. Once that is done, the same African countries will not have 'excuses' for continued environmental degradation, all in the name of development.

Furthermore, the current clarion calls for global environmental justice are based on the understanding that all human beings are coming from a position of the veil of ignorance, characterised by a hypothetical situation that ensures impartiality in the quest for global environmental justice. However, this is not the case. The environmental ethical relations between those in the global North and those in the global South are not based on equality, justice and respect for human rights. I therefore argue that one way of possibly reaching for some form of compromise is to achieve distributive justice relating to nature's benefits and burdens by appealing to the first principle of justice developed by Rawls, namely the principle of equal liberty. According to Rawls, "each person is to have an equal right to the most extensive scheme of equal basic liberties compatible with a similar scheme of liberties for others" (Rawls, 1971: 53). Following this first principle of equal liberty, all communities within the global South must be treated as having an equal right to the most extensive liberties (which could be the environment and all its benefits and burdens) compatible with similar liberties of all other persons (in this case, those within the global North, who have already developed their economies using resources from the global South). In order for environmental justice to be meaningful to all persons, nature's resources, benefits and burdens ought to be equitably and fairly available to everyone. For as long as human rights issues, injustice and inequalities between the global North and the global South remain, it would be difficult to conceive a plausible environmental justice framework in Africa.

However, this envisaged equality within the original position of humans as envisioned by Rawls could be interpreted to mean similar interests, which is not the case between African communities and those in the global North, whose relations are characterised by inequalities and varied interests. Hence, the appeal to Rawls' principle of equal liberty might enable us to appreciate that levels of development that other economies have already reached need to be matched by those which have not reached them. This is the only way in which Rawls' notion of fair shares in the distribution of benefits and burdens could be properly interpreted with reference to the conditions existing between the communities in the global North and those in the global South. One way in which this could be practically achieved is to ensure that basic natural resources for human livelihood and greenhouse emission levels between the global South and the global North should be fairly and equitably distributed as well. This is why Kelbessa proposes that "technological societies whose current emissions exceed their fair share of emissions ought to give attention to justice, and play their respective role in averting the most extreme effects" (Kelbessa, 2015: 42). However difficult, this remains one of the best ways in which the global North might think about how to contribute towards environmental justice in the global South.

From the above discussion, it is apparent that despite efforts to understand human beings as being equal to each other, inequalities persist amongst them. In this instance, inequalities in terms of the distribution of environmental benefits and burdens still exist between the global North and the global South, having been caused mainly by the operations of those in the global North on African recourses as already argued here. However, following Rawls' second principle known as the 'difference principle', the other alternative distributive framework for reaching at equality and environmental justice would be to consider developing the economies in the global South so that they match those of their counterparts in the global North. Following Rawls' second principle of difference, "social and economic inequalities are to be arranged so that they are both (a) reasonably expected to be to everyone's advantage and (b) attached to positions and offices open to all" (Rawls, 1971: 53). According to this view, if environmental justice is to be achieved in a way that is fair to all, there is need to take into consideration the conditions of the least advantaged communities, most of which are located in Africa and the global South.

Although I do not wish to get into the complex issues around African communitarianism in this particular chapter as they will be considered in the other chapters, I also take Rawls' principles of justice to be compatible with the African communitarian approach to environmental justice. Rawls's second principle of difference, for example, and its import to environmental justice lay emphasis on the distribution of needs and interests for the good of the community, and not the individual. This is what makes it easier to make a transition from an imagination of Rawls's principles of justice to those implicit in African communitarian philosophy. According to Gyekye, for example, "the notions of *sharing one another's fate, common assets, collective assets, common benefit, participating in one another's nature* – these notions and others related to them in Rawls' scheme will surely find a *more* ready embrace in the communitarian home than in the home artificially and instrumentally constructed by individuals in pursuit of their own egoistic advantages or ends" (Gyekye, 1992: 118). Similarly, if the quest for global environmental justice is understood by first appreciating these communitarian principles of justice together with the principles of equality and difference, it might be possible to find an acceptable view of environmental justice that is most suitable for the global South.

2.7 Conclusion

Arriving at an acceptable view of environmental justice involves addressing multiple social, political and economic issues. Once some of these concerns are addressed, then it might be possible to realistically think about how to equitably distribute environmental benefits and burdens and involve everyone in formulating and implementing sound environmental policies that are acceptable to all. I have attempted to highlight some of the central concerns that need to be addressed in order to formulate an acceptable global environmental justice paradigm that takes account of the realities of communities in the global South. Overall, I argue for the normative distribution of environmental justice based on Rawls' principles of equal liberty and the difference principle. Ultimately, I conclude that these two principles of justice could form the basis for the distribution of environmental benefits and burdens in an equitable manner that is acceptable to all within the global South.

References

Aristotle, (2001). Politics. (Trans. Benjamin Jowett.) In Richard MacKeon (Ed.) *The Basic Works of Aristotle*. New York: The Modern Library. 1127–1324.

Aristotle, (2011). *Nicomachean Ethics* (Trans. Robert C, Bartlett and Susan D. Collins. Chicago: Chicago University Press.

Bullard, R. D. (1990). *Dumping in Dixie: Race, Class and Environmental Quality*. USA: Westview Press.

Chemhuru, M. 2016. *The Import of African Ontology for Environmental Ethics*. D Litt et Phil (Philosophy) [Unpublished]: University of Johannesburg. Retrieved from: https://ujcontent.uj.ac.za/vital/manager/index?site_name=Research%20Output (Accessed 11 July 2018).

Chemhuru, M. (2019). Interpreting Ecofeminist Environmentalism in African Communitarian Philosophy and *Ubuntu*: An Alternative to Anthropocentrism. *Philosophical Papers*. 48 (2): 241–264.

Chemhuru, M. (2021). Land Reform and Redistribution as Environmental Justice Frameworks for Post-Colonial Africa. In, Erasmus Masitera (Ed.) *Philosophical Perspectives on Land Reform in Southern Africa*. Cham: Palgrave Macmillan, 225–240.

Chimakonam, J. O. (Ed.) (2018). *African Philosophy and Environmental Conservation*. London: Routledge.

Environmental Protection Agency, (2008). *Environmental Justice*. Downloaded from: https://www.epa.gov/environmentaljustice. (Accessed 15 October 2020).

Figueroa, R. and Mills, C. (2001). Environmental Justice. In, Dale Jamieson (Ed.) *A Companion to Environmental Philosophy*. Malden: Blackwell Publishers, 426–438.

Gyekye, K. (1992). Person and Community in African Thought. In, Kwasi Wiredu and Kwame Gyekye (Eds.) *Person and Community: Ghanaian Philosophical Studies I*. Washington D.C: The Council for Research in Values and Philosophy, 101–122.

Gonzalez, C. G. (2015). Environmental Justice, Human Rights, and the Global South. *Santa Clara Journal of International Law.* 151: 150–196.

Horsthemke, K. (2015). *Animals and African Ethics.* New York: Pelgrave Macmillan.

Ikuenobe, P. A. (2014). Traditional African Environmental Ethics and Colonial Legacy. *International Journal of Philosophy and Theology.* 2 (4): 1–21.

Kelbessa, W. (2015). Climate Ethics and Policy in Africa. *Thought and Practice: A Journal of the Philosophical Society of Kenya.* 7 (2): 41–84.

Lawson, B. E. (2008). The Value of Environmental Justice. *Environmental Justice.* 1 (3): 1–3.

Masitera, E. (2018). Economic Rights in African Communitarian Discourse. *Theoria: A Journal of Social and Political Theory.* 157 (December): 15–34.

Mawondo, S. (2008). In Search of Social Justice: Reconciliation and the Land Question in Zimbabwe. In, David Kaulemu (Ed.) *The Struggle after the Struggle: Zimbabwean Philosophical Study, 1.* Washington D.C: The Council for Research in Values and Philosophy, 7–19.

Mill, J. S. (1859/2014). *On Liberty.* London: Hackett Publishing Company.

Minority Rights Group International, (2008). *The Impact of Climate Change on Minorities and Indigenous Peoples.* Downloaded from: http://minorityrights.org/wp-content/uploads/old-site-downloads/download-524-The-Impact-of-Climate-Change-on-Minorities-and-Indigenous-Peoples.pdf (Accessed 15 October 2020).

Murove, M. F. (2004). An African Commitment to Ecological Conservation: The Shona Concept of *Ukama* and *Ubuntu*. *The Mankind Quarterly.* 45 (2): 195–215.

Murove, M. F. (2009). An African Ethics Based on the Concept of *Ukama* and *Ubuntu*. In, Munyaradzi F. Murove (Ed.) *African Ethics: An Anthology of Comparative and Applied Ethics.* Pietermaritzburg: University of KwaZulu-Natal Press, 315–331.

Nkrumah, K. (1965). *Neo-Colonialism: The Last Stage of Imperialism.* New York: International Publishers.

Plato. (1997). *Republic.* (Trans. John Llewelyn Davies and David James Vaughan). Hertfordshire: Wordsworth.

Purdy, J. (2018). The Long Environmental Justice Movement. *Ecology Law Quarterly.* 44 (4): 809–864.

Rawls, J. (1971). *A Theory of Justice.* Cambridge: Harvard University Press.

Sandel, M. J. (2009). *Justice: What's the Right Thing to Do?* New York: Farrar, Straus and GirocDonaldux.

Sayegh, A. G. (2018). Climate Justice after Paris: A Normative Framework. *Journal of Global Ethics.* 13 (3): 344–365.

Sen, A. (2009). *The Idea of Justice.* Cambridge: Belknap Press.

Tangwa, G. B. (2004). Some African Reflection on Biomedical and Environmental Ethics. In, Kwasi Wiredu (Ed.) *A Companion to African Philosophy.* Malden: Blackwell Publishers, 387–395.

3 Environmental Justice from an African Land Ethic

3.1 Introduction

The question of what an African land ethic ought to look like, and how it might contribute to environmental justice is quite crucial. It is worth noting to begin with the observation that conceptions of the land ethic in Africa continue to be informed by "skewed colonial land distortions...premised on Western entitlement or rights-oriented thought" (Masitera, 2020: 35). For this reason, any meaningful discussion around issues of the African land ethic ought to deal with fundamental environmental justice questions, especially injustice emanating from the colonial period in Africa (Conradie, 2019: 127). The following observation by Leopold remains applicable, therefore, that "there is yet no ethic dealing with man's relation to land and to the animals and plants which grow upon it. Land, like Odysseus' slave-girls, is still property. The land is still strictly economic, entailing privileges but not obligations" (Leopold, 1949: 203). However broad this view by Leopold is, its applicability to the African environment in particular could be attributed to non-African distributive patterns of land (Masitera, 2021: 1) and views of land ethic that, for the most part, continue to contribute to environmental injustice in Africa. A quest for the land ethic as the basis for environmental justice in African philosophy therefore becomes a noble exercise.

In this chapter, I consider how environmental justice could be approached following the African conception of the land ethic. Related to this, I examine the question of whether the current environmental problems responsible for environmental injustice in Africa could indeed be attributed to lack of a land ethic in African philosophy. To this end, I analyse the pre-colonial, colonial and post-colonial African conceptions of the environment. I expose how the land ethic developed and used in each of these epochs has fared in light of environmental problems and issues of injustice "related to matters of conquest, ownership, restitution, and distribution" (Conradie, 2019: 127). Having observed that environmental injustice in Africa is not necessarily attributed to the African land ethic per se, but rather to other social, political and economic issues related with colonialism, I end by examining how far

DOI: 10.4324/9781003176718-4

an African view of the land ethic might go towards contributing to conceptions of environmental justice in African philosophy.

Most views on the land ethic are in agreement on the need to enlarge the boundaries of the community so that it incorporates the environment, consisting of soils, the waters, plants and animals, etc. Within the Western philosophical tradition, for example, Aldo Leopold (1949), John Baird Callicott (1989; 2001; 2002) and Holmes Rolston (2000) propose the *land ethic* as one of the alternative views to anthropocentric environmental ethical thinking. Although they sound somewhat anthropocentric, they seek to consider the pedigree of the land ethic in terms of how it could possibly respond to "more familiar moral concerns" such as human happiness, human dignity and human rights" (Callicott, 2001: 204). Nevertheless, their views remain fundamental in conceptualising views about how to incorporate all members of the land community. A similar view is also prevalent in African philosophy. However, the significance of the land ethic, in terms of how it could respond to more familiar and current issues of environmental (in)justice in African philosophy such as land use, ownership and poverty, has not been given serious consideration. Although such a view is not new, there seems to be lack of an agreement in terms of coming up with an African land ethic that could respond to environmental injustice issues. I will argue that African ontology offers an alternative and comprehensive view of the land ethic when compared to similar views in Western philosophy. The African ontology-based view of the land ethic has not been given serious attention because of the historical circumstances of African environmental ethics and philosophy, which I have already highlighted in the first chapter. I will argue that the African view of the land ethic is much more plausible because of the way it appeals to various conceptions of existence in order to ground environmental justice.

It must be emphasised that the African land ethic, based on a pre-colonial land conception, is indeed, "an attractive and underexplored African view, and that that has been deliberately ignored" (Masitera, 2020: 37). In order to bring out my conception of an African land ethic and how it could possibly form the basis for environmental justice, I appeal to the nature of African ontology. I will use African views about ontology, i.e. – views about what it means to exist as a person (Ramose, 2003: 270), in order to construct an African ontology-based land ethic. According to this view of ontology, for example, "reality is a holistic community of mutually reinforcing natural life forces consisting of human communities (families, villages, nations and humanity), spirits, gods, deities, stones, sand, mountains, rivers, plants, and animals" (Ikuenobe, 2014: 2). The holistic, all-inclusive and non-anthropocentric character of this view makes it better positioned to form the basis for conceptualising environmental justice in African philosophy.

This chapter is divided into four sections as follows: In the first section, I give a broader conception of what the land ethic entails. In this section, I focus mainly on some of the most influential broad conceptions of the land ethic, especially as influenced by Aldo Leopold (1949), John Baird Callicott

(1989; 2001; 2002) and Holmes Rolston (2000). In the second section, I engage with the African land ethic from a historical perspective, examining the nature of land ethic in three different epochs in African environmental history, namely: the pre-colonial, the colonial and the post-colonial. In this section, I analyse the characteristics of the land ethic in each of these eras, and the implications for environmental justice. In the third section, I propose what I view to be an ontology-based view of environmental justice stemming from African philosophy. In the last section, I consider the implications of the African ontology-based view of the land ethic for a conception of property, especially how it views the land as a form common, rather that private property. In a general sense, all these sections hang together in order to demonstrate how the African land ethic is oriented towards a conception of African environmental justice.

3.2 Understanding the Land Ethic

As already mentioned, the possibility of approaching environmental justice through a land ethic has never really been given serious consideration, especially in African philosophy. In this section, I give an outline of the philosophical outlook of the land ethic in general, since I will appeal primarily to the land ethic as a point of entry and theoretical framework for environmental justice in African philosophy. Moreover, my appeal to the land ethic is mainly informed by the fact that it remains one of the most elaborate ecocentric views developed in environmental ethics so far, at least in Western philosophy, because of its history of documentation. By looking at the land ethic as ecocentric, I mean that such a view of ethical thinking is life-centred and that it takes into consideration all aspects of the ecosystem such as non-human animals, plants and other organisms etc. (Elliot, 1993: 287). Following such a perspective, I will appeal to some of the fundamental premises and conclusions within the broader view of the land ethic developed by mainly by Leopold (1949) and later popularised by John Baird Callicott (1989; 2001; 2002) and Holmes Rolston (2000). Because my focus later in the chapter will be at considering an African perspective, I take these Western views as useful theoretical frameworks and critiques on which to make some comparative analysis with African perspectives as I conceptualise the nature of an African land ethic. By and large, I will take the land ethic concept as an analytic tool to formulate my conception of an African land ethic because "an ecocentric approach such as the land ethic is not too removed from the ecological wisdom embedded in more traditional rural villages across the African continent" (Conradie, 2019: 127). For that reason, it would be important to provide a broad outline of the land ethic first before venturing into the nature of an African land ethic and its import to environmental justice.

To begin with, the land ethic is based mainly on an attempt to expand the meaning of the community to include not only human beings but also various other aspects of the land community (Earth) (Rolston, 2000: 1045) such as soils, water bodies, plant species and animals (Oruka, 1997: 244). According to Leopold, "the land ethic simply enlarges the boundaries of the community to include soils, waters, plants and animals, or collectively: the land" (Leopold, 1949: 204). Rolston also sees the land ethic as a new 'Earth ethic', which is "concerned with enlarging traditional ideas about what is of moral concern…" (Rolston, 2000: 1046). These views suggest that the land ethic might roughly be taken to be a broadening of the scope of moral considerability beyond the human community to include the entire ecosystem, which is denoted by the *land*. As I will show later on in this chapter and in the fourth chapter, such an approach is implicit within the African communitarian context, where the ethical community is not only limited to human beings but also inclusive of other non-human beings and the natural environment at large. Although Leopold is not really focused on the African communitarian view in particular, he also makes reference to this communitarian dimension to the land ethic. For him, the concept of land as "community" ought to "penetrate our intellectual life" (Leopold, 1949: 207). From this view, Leopold at least has an intention to expand the community beyond human beings using the land ethic. However, the community that he has in mind has more to do with the biotic community, while the African view which I consider is a broader view than this one.

Having highlighted their intention to expand the community beyond the human community using the land ethic, Leopold (1949) and Callicott (1989) propose a kind of biocentric environmental ethics based on the need for communitarian living and interdependence amongst those biotic parts of the land community. A biocentric view of environmental ethic is centered on "respect for life, rather than exclusively centering on humans" (Rolston, 2000: 10047). For Leopold, "a thing is right when it tends to preserve the integrity, stability, and beauty of the biotic community. It is wrong when it tends otherwise" (Leopold, 1949: 224–25). This view of the land ethics suggests that the measure of right or wrong is the extent to which an action affects the biotic community. In the same way, Callicott, in his defence of *The Conceptual Foundations of the Land Ethic*, notes that, "indeed, as the Land Ethic develops, the focus of moral concern shifts gradually away from plants, animals, soils, and waters severally to the biotic community collectively" (Callicott, 1989: 83). Understood this way, the land ethic upholds what Henry Odera Oruka (1997: 244) sees as a holistic view of ethical considerability, as it takes the biotic community and the ecosystem as ecological wholes rather than individual biological beings (Elliot, 1993: 289). According to Oruka, following the land ethic, "we have one ethics, one morality for all inhabitants of the globe, plus the globe itself. So an action or a thing is ethically right when it tends to preserve the integrity, beauty and stability of the biotic community; it is wrong when it tends to do otherwise" (Oruka, 1997: 244). From this view, the holistic and ecocentric character of the land ethic cannot be doubted. Nevertheless, its emphasis on the biotic community cannot be easily dismissed as a somewhat limited view of environmental ethical thinking. Perhaps what needs to be interrogated first is the possible influence of this biocentric slant within the land ethic. One plausible answer to this is Callicott's reading of Leopold's land ethic as drawing much from Charles Darwin (1809–1872)'s classical evolutionary account of ethics in *The Descent of Man* (1871) (Callicott, 2001: 205). According to Callicott, "what Leopold did to cook the land ethic was to take over Darwin's recipe for the origin and development of ethics, and add an ecological ingredient, the Eltonian *community concept*" (Callicott, 2001: 208). Such a biocentric conception of environmental ethics is still oriented towards anthropocentrism in so far as the notion of community revolves around the human being and his/her community.

In addition to the above biocentric view, which in itself could be read as a limited view of environmental ethics, the land ethic proposed by Leopold and Callicott is oriented towards some sort of ecological communitarianism. According to Leopold, "all ethics so far evolved rest upon a single premise: that the individual is a member of a community of interdependent parts. His instincts prompt him to compete for his place in that community, but his ethics prompt him also to co-operate (perhaps in order that there may be a place to compete for)" (Leopold, 1949: 203–204). Understood correctly, this view of the land ethic tries to bring together individual members of the community into some sort of communitarian relationships. However, the only problem with such relationships is that it is mainly based on interdependence and mutual benefits only for interested parties who at the same time compete for their survival in the same community. Leopold further demonstrates this view through a hierarchy of beings that can be conceptualised in the *land pyramid* as follows: "the bottom layer is the soil. A plant layer rests on the soil, an insect layer on the plants, a bird and rodent layer on the insects, and so on up through various animal groups to the apex layer, which consists of the larger carnivores" (Leopold, 1949: 215". This hierarchy might be understood as legitimising an anthropocentric view because at the apex of the hierarchy are human beings. In addition to this anthropocentric pyramid, Leopold argues that only human persons have "ecological conscience, and this in turn reflects a conviction of individual responsibility for the health of the land" (Leopold, 1949: 221). Thus, despite the attempt in such a view at concern with the 'health of the land', the fact that it is still rooted in a sort of anthropocentric epistemology makes this understanding of the land ethic objectionable.

Understood from the above-mentioned perspectives, the land ethic attempts to be holistic and non-anthropocentric. However, it remains fairly anthropocentric, as Callicott opines that "the land ethic … has a holistic as well as an individualistic cast" (Callicott, 1989: 83). This is because the reasons proffered for the expansion of the understanding on the 'community' are not exclusively motivated by the need to safeguard the biotic community and avoiding the ecological crisis. Such reasons are also motivated by the need for continued interdependence by the various beings within the land community, especially conscious beings, and human beings in particular. This view is anthropocentric because it does not consider the land community for its own sake, but somehow for continued living and posterity of humans and their biological counterparts. As a result, Rolston sees Leopold's land ethic as having managed to "blend anthropocentric and biocentric values" (Rolston, 2000: 1045). In spite of these possible objections, however, the land ethic remains an important theoretical framework for conceptualising environmental justice between and among the various beings that constitute the land community because of its emphasis on aspects such as the biological community, interdependence and ecological holism.

The land ethic could be seen as going further than the economic and instrumental view of beings because of its view of the biotic community as having value. Leopold insists that such a value "is broader than mere economic value" (Leopold, 1949: 223) while Callicott takes it as "intrinsic value" (Callicott, 2002: 10). Nevertheless, the philosophical pedigree of such value and the land ethic itself are not so clearly explained over and above the appeal to biology. This explains why Rolston (2000: 1045) envisions an Earth ethic beyond the land ethic. In it, he tries to go beyond a life-centred view of environmental ethics and justice. I therefore see the African land ethic as an alternative view of environmental ethics and justice because of its ontological underpinnings and the way it attempts to go beyond living reality in the quest for environmental justice in African philosophy, as I will proceed to show.

By and large, the attempt to understand the land ethic from ecological communitarianism, holistic, non-anthropocentric and non-economic perspectives may be taken as ethically plausible. Although Oruka is looking at the land ethic broadly, he notes that, "the land ethic, as a proposal for what should embody 'environmental ethics', treats all inhabitants of the earth plus the earth itself as holy or valuable as human beings are" (Oruka, 1997: 244). It therefore a fact that the land ethic places much emphasis on such important aspects such as community, ethical holism and non-anthropocentrism, which could all be used to interpret conceptions of environmental ethics and justice in African philosophy. Having examined these perspectives based on the land ethic, I seek to proceed and consider some of the underexplored views of environmental justice based on the African ontological conception of existence and the land ethic in the following sections. First, I will take a brief historical perspective and look at how the African land ethic itself has evolved over time.

3.3 The African Land Ethic: A Historical Overview

In this section, I consider the African land ethic from a historical perspective. I look at how the pre-colonial, the colonial and post-colonial periods could have shaped conceptions of the African land ethic. I then consider what the implications of such different land ethic in these periods could have been for environmental justice issues such as land ownership, access to land and natural resources, inequalities and poverty. I highlight the different kinds of land ethic persisting in each of these different epochs and their implications for environmental justice. I end by proposing why there is need to appeal to the African ontology-based land ethic that is characteristic of the pre-colonial period as opposed to the economic and largely anthropocentric models of the land ethic based on looking at the land from a private property perspective.

While I present my conception of the African land ethic as being traceable to the pre-colonial era, I do not wish to pretend that pre-colonial African communities did not have problems associated with environmental (in)justice. However, I must emphasise that despite the existence of environmental injustice concerns, which may not be as pronounced as what is the case now, such communities had a systematic conception of the land ethic based on their ontological connections with the environment, particularly the land. Ikuenobe echoes the same sentiment about the existence of a land ethic in pre-colonial African societies, and emphasises that environmental problems in Africa "did not exist prior to colonialism because traditional Africans had conservationist values, moral attitudes, practices and way of life" (Ikuenobe, 2014: 1). Moreover, this attitude is complemented by the absence of the economics-based model of the land ethic in African philosophy. The pre-colonial African land ethic was mainly anchored on the idea of belonging to the land, rather than owning the land (Conradie, 2019: 129). The individual belonged to the land because of the ontological connections s/he had with it, together with other various communities of the land. In terms of environmental justice, this membership view of the land ethic made it possible and unproblematic for these traditional African communities to communally 'allocate' each other elements of the environment such as the land and other natural resources. Although my use of the word 'allocate' here might be understood from an anthropocentric view, it needs to be emphasised that such 'allocations' or distributions were not based on the economic model of ownership that is prevalent in non-African traditions. These kind of 'allocations' were done on the basis of belonging to the land, and having various relationships of interdependence between members of that land community.

With the advent of colonialism in Africa in the late 19th century, the African land ethic, which was based on belonging, communal ownership and ontological connections between the land and the community, was abandoned. Yet, prior to colonialism, the African ontology-based land ethic was the mainstay of African "traditional conservationist values, ways of life, and moral attitudes… [before they] were destroyed by the exploitative ethos of European colonialism and modernity" (Ikuenobe, 2014: 1). As a result, the economics-based model of the land ethic effectively replaced the African land ethic, although not completely as it is still inherent in African philosophy itself. Literally, the African land ethic could no longer work effectively because of the dispossession, land imbalances and poverty, among a host of other environmental injustice issues confronting the African populace after colonialism (Chemhuru, 2021: 227). As Mawondo sees it, "neglecting the fact that black poverty in [Africa] is to a large extent due to the concrete historical acts of deliberate dispossession is thus a continued act of violence against them, hence an injustice" (Mawondo, 2008: 9). Accordingly, a colonial land ethic in Africa could be viewed as one of the anchors of environmental injustice in Africa, and it could be said that such a land ethic ought to be deconstructed in order to solve the problems of "landlessness" and "inequalities" (Mawondo, 2008: 9) in post-colonial Africa. For this reason, it would be worth taking seriously, my submission elsewhere that "land reform and redistribution are necessary frameworks for post-colonial Africa's quest for equality, justice as well as socio-economic development through agrarian reform" (Chemhuru, 2021: 225). In the same vein, the African land ethic would be incomplete without a viable agrarian structure, which should address landlessness and inequalities that are responsible for environmental injustice in post-colonial Africa.

As I emphasise throughout this work, the African post-colonial condition continues to be strongly marked by its history and conditions that were shaped by colonialism. By way of example, most African communities today are characterised by a lack of access to land and mineral rights, marginality, informal settlements and poverty. Conradie avers that "there are endless debates on the impact of nature conservation on indigenous peoples, land use rights, mineral rights, informal settlements and so forth" (Conradie, 2019: 128). These are environmental justice issues that could be attributed to the colonial land ethic because "colonial structures left behind still continue to engender and contribute to environmental problems in Africa today" (Ikuenobe, 2014: 1) and to environmental injustice in particular. For this reason, Sam Moyo feels that "the enduring challenge is to redress colonially driven and post-independent unequal land ownership, discriminatory land use regulations and insecure land tenure systems which marginalise the majority of rural and urban populations" (Moyo, 2007: 60). From this view, it is even worse to imagine that environmental injustice issues to do with access to land-use and ownership are mostly felt by African indigenous populations that were dispossessed and marginalised by these colonial powers. Ultimately, poverty, which is one of the serious environmental justice issues in Africa is the order of the day for such communities that lack access to basics such as the land. However, for Mawondo, "while the mere fact of poverty is not necessarily indicative of the existence of injustice, the existence of poverty due to deliberate acts of dispossession and marginalisation is evidence of injustice" (Mawondo, 2008: 10) and in particular, environmental injustice affecting poor African indigenous populations. One could therefore argue, and for good reasons, that much of the poverty in post-colonial Africa could be directly attributable to environmental injustice.

Against this historical background, the reality of environmental injustice in Africa owing to colonial and neo-colonial land ethic models cannot be denied. This is why I maintain that the way to approach environmental (in) justice in post-colonial Africa would be to firstly abandon the non-African economics-based models of the land ethic and consider how conceptions of the traditional African land ethic prior to colonialism might be useful. In this regard, I adopt the African ontology-based view of the land ethic as an alternative. In the following section, I therefore focus on the nature of such an African land ethic, which is rooted in African conceptions of existence, and how it might be considered as the basis for environmental justice in African philosophy.

3.4 The African Land Ethic from an Ontological Perspective

The hermeneutics of an African land ethic cannot be undertaken without making reference to African ontology. Loosely, African ontology could be understood as roughly concerned with the ideas about existence or being in general, especially categories in the order of primogeniture within such beings. According to African ethics, the existence and order of beings in African ontology will ultimately determine the ethical considerability of the various related beings in African ontology. Similar interpretations about how African ontology undergirds other fundamental areas such as African ethics and epistemology have been done by Onyewuenyi (1976), Gyekye (2002; 2007; 2010) and Ikuenobe (2014). Gyekye (2010: 102) for example, challenges us to consider how the ontological understanding of the individual and the community in Africa could be taken as the basis for understanding African normative issues. In the same vein, in this section I will show how African ontologies inform what an African land ethic ought to be like. Specifically, I will consider how the ontology-based African land ethic broadens the understanding of the land community from the biotic community to include all beings, whether biotic or non-biotic, and future generations.

The African idea of ontology, as I have alluded to above, deals with notion of being or existence which could at least be rationally understood, unlike metaphysics which is beyond human rational capacity (Poli, 2010: 1). At the same time, that notion of being or existence should not be understood as only limited to human beings and their existence alone. In fact, the idea of being is also rationally applicable to the surroundings or the environment because the environment itself can at least be partially understood by human beings. Onyewuenyi examines African ontology and proffers a definition that is closer to this view as he sees ontology as "the science of *being as such, the reality that is*" (Onyewuenyi, 1976: 524). This implies that there is no distinction between humanity and nature. From this view, the idea of land cannot be dissociated from African ontology because both, the notion of land and African ontology are part of the same being or reality. Rakotsoane also looks at African ontology and submits the view that "Africans maintain a monist world-view" (Rakotsoane, 2005: 9) in which reality is largely compact and somewhat unified. The African land ethic is therefore informed by such an African conception of a monist world-view or *being,* where, despite the existence of various beings, there is some sort of unity in the order of such existence. From this perspective, the African land ethic is therefore characterised by "a single, integrated totality of cause and effect, in which different components (including man) hang harmoniously together like threads of a spider's web" (Rakotsoane, 2005: 9). This view points to a monist but unified and all-inclusive web of being within the African land ethic. According to such a perspective, although different beings constitute the web of existence, they all are unified towards life and well-being such that despite the interdependence among them, they are somewhat united towards the preservation of the land to which they find life. If taken seriously, such an understanding of beings could be helpful in constructing a plausible conception of environmental justice because of its emphasis on the land, or reality as unified.

In African ontology, human existence cannot be distinguished from the idea of the land. In such an ontology, there are thought to be some close spiritual and ontological connections between the idea of the land and the notion of existence. The expression a *son/daughter of the soil* is usually used with reference to native Africans when they emphasise their ontological connectedness with the land or environment at large. In an African child poem titled, *African Child,* Eku McGred highlights the importance of this expression for posterity as follows:

> I am the son, daughter of the soil
> Rich in texture and content
> Full of potential for a better tomorrow
> Teach me discipline, teach me character, teach me hard work
> … I am an African child (McGred, 2010).

The expression *son/daughter of the soil,* which is emphasised in this poem is often used in attempts to highlight the need to guard against one's inhabitant – the land, source of livelihood and natural resources, and especially against foreign invasion, colonial and neo-colonial machinations. However, the expression also goes as far as showing the connections between human beings, future generations and their land. For that reason, it has very strong implications not only for patriotic reasons but for environmental justice as well. This is because it encourages a strong connection between human beings and the land or the surrounding environment at large, such that human beings claiming to be sons and daughters of the soil will always guard against it and its inhabitants.

Another view that also confirms the ontological interconnections between human beings and the land/environment is reflected primarily in the reluctance by most Africans to be resettled to a new environment. When, for example, governments identify risk factors such as proximity to industrial effluent, low-lying and flood-prone areas, or when they come up with new projects like urban expansion, water projects and mining that demands that some people should be resettled elsewhere, there is a general reluctance among the African indigenous populace to relocate. In most cases, no matter how attractive the resettlement programme might be, these affected individuals usually cite the ontological link between the land and the people. Writing in *The Conversation*, Kenneth Tafira argues that "…African personhood and *being* revolve around earth and all that walks on it, the heavens, the waters and all that live in it, the natural landscape, the atmosphere and livestock" (Tafira, 2015). The implications of Tafira's argument here are that it reflects how the idea of African existence in general and the African land ethic are closely knit to the extent that there is ecological harmony in this ontological view of reality.

It must also be emphasised that the African land ethic is largely based on a unitary view of being or existence. According to this view, reality is construed as a unity in which the environment is considered to be a whole without separating it from any of its constituent parts like the rational human beings, the biotic and the non-biotic community. This non-anthropocentric view resonates well with the African ontological view of existence, in which reality is considered to be an aspect of the same existence. Jimoh sees the "African ontological notion of reality as a continuum in which both the subject, as the cognitive agent, and the object, as the cognized phenomenon, are part and parcel of the same reality" (Jimoh, 2017: 121). In other words, the understanding of either one aspect (the land) or the other (human existence) cannot be complete without the other. In effect, these two (land or human existence) complement each other. While ontology is generally taken and used with reference to human existence, the land is also taken as being part of that existence. This view is aptly captured by Verhoef and Rathbone as follows:

> Land is thus embedded in the concept of ontology and it cannot be separated from it. In other words, land is constituted by one's ontology, but on the other hand one's ontology is constituted by land (a 'hermeneutical cycle'). An ontology of the land is thus the constituting aspect that contributes to one's understanding of the reality of land and how one's *land* constitutes one's reality (Verhoef and Rathbone, 2015: 156).

Implicit in the above assertion is the idea that the African land ethic is also central to, and inseparable from, human existence. This is contrary to an anthropocentric view that is implicit within Western philosophy, which is based on "a dualism of man and nature [and] insists that it is God's will that man exploit nature for his proper ends" (White, 1967: 4). In contrast, according to the African ontological view, land is closely linked to one's ontological existence and identity. The implications of this for environmental justice are that, the basis for considering environmental justice issues with regards to human relationships with other human beings is that humans are not only part of the same ontology and related to each other but also that the land itself is also part of their ontological reality. For this reason, they are somewhat *related* to the land.

There is some sense of responsibility and accountability for environmental justice that comes from the realisation that the land is part of one's ontology.

Within the African land ethic, all beings, including human beings make up the conception or essence of what the land is. By this, I mean that human beings and other aspects of the environment ultimately define what the land is by ontologically and collectively composing it. Accordingly, such a land ethic is incomplete if the human persons are not part of it. As I have argued above, an African land ethics is fundamentally different from an anthropocentric land ethic, which is essentially based on separating the human being from the environment on the basis of human reason. The story of creation in the Judeo-Christian heritage supports this perspective, as it accepts the dominance of humans over other creations (White, 1967: 4). From African ontology-based land ethic, however, these humans and the environment ought not to be separated from the rest of reality because they collectively define what the essence of reality is. Ikuenobe captures this view as follows: "in the Euro-American worldview, there is a separation between the self and the non-self (phenomenal world). Through this separation, the phenomenal world becomes an Object, an *it*... The phenomenal world becomes an entity considered as totally independent of the self" (Ikuenobe, 2014: 3). This understanding of reality in Western thinking is reflected in the way philosophers such as Newton, Descartes, Hume, Locke and Kant (Ikuenobe, 2014: 3), in their different epistemic approaches, place so much emphasis on how humans are separate from the world of knowledge. It also explains why in this regard, Locke went even further than these perspectives by advocating for the functionalist view of the land in his defence of private ownership of property based on the "labour" of one's body and the "work" of one's hands. (Locke, 1823: 115). The African land ethic is not, however, compatible with such a functionalist and economics-based view of the land as I will show in the African conception of property in the next section. This African land ethic does not see a divide between the human being and the land, nor some commercial value in it, and the implications of such a view are that it instils a sense of responsibility and accountability in human beings as they relate with the land and other human and non-human beings, as they are always conscious of the fact that the land is part of their ontology.

Within the African land ethic, the duties to environmental justice are also thought to be transcendental to the network of existence. By way of example, the African ontology-based land ethic is based on the inseparability of the land from the various forms of being in African ontology such as the spiritual beings, humans, the dead and the future generations. As Tafira puts it, the African "land belongs to the living, the dead and the unborn, making it inalienable" (Tafira, 2015). Tafira could be interpreted as saying that land is an inseparable aspect of African ontological network, which stretches from the spiritual beings, the dead, the living and future generations (see also, Ikuenobe, 2014: 2). The implications of this view of the land ethic are that the land is treated not only as the habitat of the biotic community like we have seen in the Western view of the land ethic, but that it is taken as belonging to various beings including spiritual beings and the future generations.

If this view is properly understood, it means that lack of access to land and land dispossessions (i.e. experienced by the majority of African populations during the colonial era) have serious implications not only for environmental justice (Chemhuru, 2021) but also for questions of being and identity.

Another key feature of the African land ethic is its emphasis on respect for all existence, which is based on reverence for the land. The land is revered for various spiritual reasons, some of which might be difficult to justify philosophically. However, such reverence also has some conservationist and preservationist implications for the entire ecosystem and environmental justice. The idea of revering the land is based on the understanding that the land belongs to the wider community of the spiritual, the departed, the living and the future generations (Dzobo, 1992: 231). This kind of African ontological view of reality is based on what Magesa sees as "...the sacrality of life, respect for the spiritual and mystic nature of creation, and, especially, of the human person; the sense of the family, community, solidarity and participation; and an emphasis on fecundity and sharing in life, friendship, healing and hospitality" (Magesa, 1997: 52–3). This understanding of the land and nature in general might be usefully taken as oriented towards a conception of environmental justice. This is because of its emphasis on respect for the entire ecosystem together with its other-regarding orientation.

In addition to this reverence for the land, the philosophy of *unhu/ubuntu* is another important aspect of African ontology, which is also central to the African land ethic. African ontology cannot be discussed without making reference to *unhu/ubuntu* (Samkange and Samkange, 1980; Ramose, 1999) because *unhu/ubuntu* is actually part of African ontology. Although the philosophy of *unhu/ubuntu* is dominant in Southern Africa, it has generally been accepted as philosophically plausible in much of sub-Saharan Africa. African ontology is rather inextricably anchored on this philosophy. Ramose's (1999) definition of *ubuntu,* for example, implies that it is essentially an African ontological and ethical philosophy of existence. For him, the term *ubuntu* is derived from two conjunctions – *ubu* implying the notion of *being* or existence in general and – *ntu* signifying the nodal point at which *beingness* achieves fullness or concrete form (Ramose, 1999: 49–50). What is important is this definition of *ubuntu* is not just the etymological understanding of it, but also the humanistic import which the view has for various dimensions of human life, including environmental ethics and justice. Metz and Gaie see the African approach through *ubuntu* as providing "a unitary foundation for a variety of normative and empirical conclusions" (Metz and Gaie, 2010: 273). It is therefore instructive that interpretations of the conception of *unhu/ubuntu* point to a humanistic ethic, one that has much to do with a conception of moral humaneness or the ethical existence of the person based on humanness, or what it means to be a human person first. With reference to its import for the African land ethic, the idea of *unhu/ubuntu* can be taken as responsible for instilling fundamental values about being humane and communal living. Through *unhu/ubuntu*, "the humanness and communitarian life are centred on values such as relationality, cooperation, common good, and equal distribution of communal goods" (Masitera, 2020: 35). If these virtues of *unhu/ubuntu* are properly understood, it means that a land ethic based on human relationality, cooperation, the common good and equal distribution of communal goods such as land and other natural resources would be the basis for a plausible conception of environmental justice.

It must be emphasised that African ontology-based land ethic is an all-encompassing view of environmental ethics and justice. According to Verhoef and Rathbone, "this 'encompassing ontology of land' is reflected in the theologies of creation that emphasise the centrality of land in human history as a gift from God shared by all people as human habitat" (Verhoef and Rathbone, 2015: 157). Significantly, this all-encompassing view of the land ethic has been threatened by the twin processes of colonialism and capitalism in Africa. Even within the post-colonial African context, despite the end of colonial rule, capitalist economies still persist such that the African land ethic is characteristically economics-based. To this end, Verhoef and Rathbone acknowledge the existence of what they call "a modernistic ontology of land that in general reduces land to functionalistic and cultural concepts" (Verhoef and Rathbone, 2015: 156). Although I concur with them on the view that a modernistic ontology of the land is functionalist in so far as it is largely informed by an economics-based view of the land, I do not accept their doubts about an African cultural ontology of the land or an African ontology-based view of land ethic. I will show my conviction in an African ontology-based land ethic in my response to objections that may be raised against such a land ethic below.

At this juncture, I may also need to acknowledge and earnestly respond to some objections that my appeal to an African ontology-based land ethic might face. The first objection relates to the controversial metaphysical suggestions of an African ontology, especially its accommodation of existence of reality beyond that which can be physically comprehended, at least in empirical terms. The major weakness for which an African ontological hierarchy of existence is criticised is its acknowledgement of the existence of the metaphysical realm, i.e., spiritual beings such as God and the ancestors within such a hierarchy (Dzobo, 1992; Magesa, 1997). Although this metaphysical view is not only unique to African ontology, it has always been difficult to find sufficient philosophical grounds for defending such a position. However, it remains an important aspect of ontology not only in African philosophy but also in other non-African traditions that acknowledge the existence of beings beyond the physical. Perhaps the best way to accept this kind of knowledge from African philosophy is to accept the view that "No matter how educated (in terms of western schooling) … Africans always have their unique way of explaining reality. They often understand and explain things and events in a way that appears strange, if not incomprehensible, to non-Africans" (Rakotsoane, 2005: 8), the area of African ontology and metaphysics being one of them. However, I do not intend to present a solipsist and closed view of African ontology altogether. I will therefore present such an African conception of existence in terms of its metaphysical implications to environmental justice in African philosophy without necessarily meaning that it is immune to this kind of criticism that I have raised above.

Contrary to my proposal of an African ontology-based land ethics, Verhoef and Rathbone (2015) consider the possibility of "a theologically informed ontology of the land [as capable of] providing ethical criteria to address the injustice of reductionism in the current process of land redistribution" (Verhoef and Rathbone, 2015: 157) in Africa. In their view, an African ontology-based land ethic is, of course, one of these reductionist ontologies of the land. They identify the economics based view of the land ethics and the African ontology-based land ethics and view both as reductionist ontologies as follows: "On the one hand land is understood as mere economic function…On the other hand, land is understood as that which gives people their dignity, integrity and identity, and links them to their history, ancestors and culture" (Rathbone and Verhoef, 2021: 267–8). It is worth noting that Verhoef and Rathbone (2015) seek to propose a theologically informed ontology of the land, although the African land ethic accommodates such a theological view, if it is properly understood because theology is also part of African ontology. The acknowledgement of the existence of the God and the ancestors in African ontology, for example, is a theological view. Although Rathbone and Verhoef (2021) see both ontologies (economics-based and African ontology-based) of the land as reductionist, I will adopt the latter because of its ontological appeals. Such a view would guarantee environmental justice for all beings that are part of the African hierarchy of ontology because of its appeals to integrity, dignity and identity, which Rathbone and Verhoef acknowledge to be implicit therein. These three aspects are all important to African ontology and could be meaningfully taken as significant grounds for a conception of African environmental justice.

3.5 The African Land Ethic, the Idea of Property and Environmental Justice

In the preceding section, I have shown how the African land ethic is closely tied to African ontological conceptions of existence. In this particular section, I seek to demonstrate how the same ontological views about existence and the land ethic could inform certain communitarian conceptions of the view of land by which "natural endowments are communally owned" (Masitera, 2020: 42) to the extent that they could also influence conceptions of environmental justice. Specifically, I will consider the African land ethic as a type of common property (Nyerere, 1968), making it an important part of communitarian relationships with human beings and other types of beings. Ultimately, I seek to outline the implications of such a view of the land as communal property for environmental justice.

As I have already discussed in the first chapter, the African communitarian view of existence is one of the most defining and key ontological features of African social structures (Nyerere, 1968; Mbiti, 1969). According to this perspective, notwithstanding the fact that African communities are not homogeneous, it is a fact that communitarian existence is a fundamental characteristic of most of the African communities (Menkiti, 1984; 2004; Gyekye, 2002; 2010). At the same time, the idea of communitarian philosophy has some implications for a kind of land ethic that is informed by such existence. However, such a dimension has not been examined in greater detail. As I observe here, African communitarian thinking influences a communitarian view of the land ethic, where land and natural resources belong to the community and not individual persons. Indeed, according to such an African communitarian land ethic, the land is the common property of all beings because for these African traditional communities, the "land was always recognised as belonging to the community" (Nyerere, 1968: 7). Following Nyerere's view, which is true of how the land is viewed in African philosophy, it follows that the land should therefore not be conceived as the private property of individuals. This view does not necessarily mean that individuals did not have allocations in the distributive patterns of such communitarian land, natural resources and other basic rights. As Gyekye observes, for example, "individual rights such as right to equal treatment, to *our* property, to freely associate with others, to free speech and others, would be recognised by communitarianism, especially of the restricted or moderate type" (Gyekye, 2002: 311). In this view by Gyekye, I have deliberately italicised the word 'our', which is making reference to property. I did so in order to show the extent to which Gyekye's view attempts to confine property to the community despite the distribution of such basic rights and needs of individual members of the community. Gyekye's assertion could be interpreted to imply that property still belongs to the community, even if it is distributed to individuals. This argument is also shared by Masitera, who notes that, "beyond communal ownership, communal property is also shared such that every member of the community benefits from that property" (Masitera, 2020: 42). This means that land and all natural resources should be conceived and distributed following the same communitarian framework, which seems to be essentially distributive. Nevertheless, it must be borne in mind that such shares are not meant to imply individual ownership of properties, as the community remains the ultimate guardian.

Thus far, it is clear that the communitarian view of existence points to the idea of communitarian property (Gyekye, 2002: 311). It also influences the communitarian land ethic, thereby influencing environmental justice. Moreover, the idea of value is also central in any conception of property, and ultimately, that has a bearing on the conceptions of environmental justice because not all value is moral (Metz, 2019: 29). However, within the African communitarian conceptions of the land, for example, even if property is conceived from a communitarian perspective, it also has inherent value. Although the idea of 'value' could be loosely understood as the price or worth of something, it can also be understood more deeply to mean the value or worth of something either in the economic or philosophical sense. If something has value in the economic sense, it has instrumental value, and if it has value in the philosophical sense, it has intrinsic or inherent value (O'Neill, 2001: 164). From a Kantian perspective, if a thing has value in the latter sense, which means that such a thing should be treated as an *end* in itself and not as a means to some other *end*. Accordingly, the African land ethics associates the land more with intrinsic than instrumental value because of its emphasis on the ontological connections with the land, which I have discussed in the previous section. It should be emphasised that in such a land ethic, intrinsic value differs in terms of degree from one being to the other (Metz, 2019) within that land community. Nevertheless, it is also not possible that an object could possess both intrinsic and extrinsic value (in this case, economic value) at the same time. According to the African land ethic, although the land could be conceived as having intrinsic value in so far as the various beings living in the ecosystem each have their own individual inherent value, the land does not have value in the economic sense, or that it does not have economic value. Consequently, the idea of the intrinsic value of the land in the Africa land ethic has far-reaching implications for environmental justice.

Contemporary African political institutions have abandoned the African ontology-based land ethic and adopted the economics-based view of the land as a sort of private property. According to these views, which are in contrary to the African land ethic, "property is seen to be exclusive to the individual. This is the core of capitalist relations, which are embedded in colonial modernity" (Tafira, 2015). This explains why the political gimmick that *the land is the economy, or the economy is the land*, is appealing in most post-colonial capitalist African countries. Most post-colonial African governments rely on such political and economic gimmicks to attract voters, who are mostly peasant farmers in need, or possession of land that would have been given primarily on partisan basis. However, such an approach is based on an economics approach to the land, and not only contrary to the African land ethic but also largely responsible for the environmental injustice in the form of lack of access to the land and natural resources by people who are, for the most part, poor. Following the African land ethic, land should not be conceived as a private property or a commodity in the economic sense. It ought to be taken as part of what it is to be or to exist, as I have argued in the previous section.

The economics-based conception of the land ethic also has strong implications for the ontology-based view of the land and ultimately to environmental justice. For example, most of these conceptions view the land as a form of private property such that individual persons, communities, organisations, private players, etc., can actually acquire it and either economically develop it, or in some cases speculate so that it continues to gain economic value. This is why Nyerere thinks that "the foreigner introduced a completely different concept of land as a marketable commodity. According to this system, a person could claim a piece of land as his own private property whether he intends to use it or not" Nyerere, 1968: 7). These conceptions of private property can be justified using either Aristotelean or Lockean views of private property as well as colonial conceptions of land as a commodity. According to Tafira, "the colonisers brought with them a Eurocentric perception that the land was a commodity to be purchased and sold. Of course the legal justification in the form of laws, particularly English common and Roman-Dutch laws, was to legitimise the commoditisation of land" (Tafira, 2015). Such conceptions continue to influence even contemporary post-colonial African political institutions. However, the appeal to African ontological views of the land and their import for the communitarian view of the land as common property cannot be underestimated.

Overall, if we are to meaningfully conceptualise a plausible environmental justice framework from an African land ethic, we must settle for a view of property that does not create injustice and disparities in the distributive patterns of environmental goods. Nyerere sums up this view by arguing that we must "reject the capitalist attitude of mind which colonialism brought into Africa. We must reject also the capitalist methods which go with it. One of these is individual ownership of property" (Nyerere, 1968: 7). Fortunately, the African land ethic does not accept the kind of capitalist attitude and private ownership of land and natural resources. This is why I take Nyerere's perspective and submit that the African land ethic remains one of the ways through which environmental justice might be approached in African philosophy because of the various reasons that I have submitted in this particular chapter.

3.6 Conclusion

In this chapter, I have attempted to come up with a rather underexplore, yet novel conception of environmental justice based on the African land ethic. In order to construct such a land ethic, I have appealed mostly to the African ontological conceptions of existence and views about the land ethic. I demonstrate the way in which such views could be understood as possible and plausible approaches to environmental justice in African philosophy. More specifically, I have tried to show how environmental equity among different beings within the environment might be realistically achieved using this land ethic based on African ontological conceptions of existence. It is my hope that this perspective of environmental justice will go a long way in showing how African philosophies might make a plausible contribution to environmental justice and philosophy.

References

Callicott, J. B. (1989). *In Defence of the Land Ethic: Essays in Environmental Philosophy*. New York: SUNY Press.

Callicott, J. B. (2001). The Land Ethic. In, Dale Jamieson (Ed.) *A Companion to Environmental Philosophy*. Malden: Blackwell, 204–217.

Callicott, J. B. (2002). The Pragmatic Power and Promise of Theoretical Environmental Ethics: Forging a New Discourse. *Environmental Values*. 11 (1): 3–25.

Chemhuru, M. (2021). Land Reform and Redistribution as Environmental Justice Frameworks for Post-Colonial Africa. In, Erasmus Masitera (Ed.) *Philosophical Perspectives on Land Reform in Southern Africa*. Cham: Palgrave Macmillan, 225–240.

Conradie, E. M. (2019). A (South) African Land Ethic? The Viability of an Ecocentric Approach to Environmental Ethics and Philosophy. In, Munamato Chemhuru (Ed.) *African Environmental Ethics: A Critical Reader*. Cham: Springer, 127–139.

Dzobo, N. K. (1992). Values in a Changing Society: Man, Ancestors and God. In, Kwasi Wiredu and Kwame Gyekye (Eds.) *Person and Community: Ghanaian Philosophical Studies 1*. Washington D.C: The Council for Research in Values and Philosophy, 223–240.

Elliot, R. (1993). Environmental Ethics. In, Peter Singer (Ed.) *A Companion to Ethics*. Malden: Blackwell, 284–293.

Gyekye, K. (2002). Person and Community in African Thought. In, P.H Coetzee and A.P.J. Roux (Eds.) *Philosophy from Africa: A Text With Readings*. New York: Oxford University Press, 297–312.

Gyekye, K. (2007). *Tradition and Modernity: Philosophical Reflections on the African Experience*. New York: Oxford University Press.

Gyekye, K. (2010). Person and Community in the Akan Thought. In, Kwasi Wiredu and Kwame Gyekye (Eds.) *Person and Community: Ghanaian Philosophical Studies 1*. Washington D.C: The Council for Research in Values and Philosophy, 101–122.

Ikuenobe, P. A. (2014). Traditional African Environmental Ethics and Colonial Legacy. *International Journal of Philosophy and Theology*. 2 (4): 1–20.

Jimoh, A. K. (2017). An African Theory of Knowledge. In, Isaac E. Ukpokolo (Ed.) *Themes, Issues and Problems in African Philosophy*. Cham: Palgrave Macmillan, 121–136.

Leopold, A. (1949). *A Sand County Almanac and Sketches Here and There*. Oxford: Oxford University Press.

Locke, J. (1823). *Two Treatises of Government. Retrieved from:* https://www.yorku.ca/comninel/courses/3025pdf/Locke.pdf. (Accessed 3 May 2021).

Magesa, L. (1997). *African Religion: The Moral Tradition of the Abundant Life. New York*: Orbis Books.

Masitera, E. (2020). Indigenous African Ethics and Land Redistribution. *South Africa Journal of Philosophy*. 39 (1): 35–36.

Masitera, E. (2021). Thinking About Land Reform in Southern Africa: The Introduction. In, Erasmus Masitera (Ed.) *Philosophical Perspectives on Land Reform in Southern Africa*. Cham: Palgrave Macmillan, 1–15.

Mawondo, S. (2008). In Search of Social Justice: Reconciliation and the Land Question in Zimbabwe. In, David Kaulemu, (Ed.) *The Struggle after the Struggle: A Zimbabwean Philosophical Study 1*. Washington D.C.: The Council for Research in Values and Philosophy.

McGred, E. (2010). *African Child*. Retrieved from: http://www.worshipthroughthe-storm.com/african-child-2/. (Accessed 15 March 2021).

Mbiti, J. S. (1969). *African Religions and Philosophy*. London: Heinemann.

Menkiti, I. A. (1984). Person and Community in African Traditional Thought. In, Richard Wright (Ed.) *African Philosophy: An Introduction*. Lanham: University Press of America, 171–181.

Menkiti, I. A. (2004). On the Normative Conception of a Person. In, Kwasi Wiredu (Ed.) *A Companion to African Philosophy*. Malden: Blackwell Publishers, 324–331.

Metz, T. and Gaie, J. B. R. (2010). The African Ethic of Ubuntu/Botho: Implications for Research on Morality. *Journal of Moral Education*. 39 (3): 273–290.

Metz, T. (2019). An African Theory of Moral Status: A Relational Alternative to Individualism and Holism. In, Munamato Chemhuru (Ed.) *African Environmental Ethics: A Critical Reader*. Cham: Springer, 9–27.

Moyo, S. (2007). The Land Question in Southern Africa: A Comparative Review. In, Lungisile Ntsebeza and Ruth Hall (Eds.) *The Land Question in South Africa: The Challenge of Transformation and Redistribution*. Cape Town: Human Sciences Research Council, 60–86.

Nyerere, J. K. (1968). *UJAMAA: Essays on Socialism*. Dar es Salaam: Oxford University Press.

O'Neill, J. (2001). Meta-Ethics. In, Dale Jamieson (Ed.) *A Companion to Environmental Philosophy*. Malden: Blackwell Publishers, 163–176.

Onyewuenyi, I. (1976). Is There an African Philosophy? *Journal of African Studies*. 3 (4): 513–528.

Oruka, H. O. (1997). *Practical Philosophy: In Search of an Ethical Minimum*. Nairobi: East African Educational Publishers.

Poli, R. (2010). Ontology: The Categorical Stance. In, Roberto Poli and Johanna Seibt (Eds.) *Theory and Applications of Ontology: Philosophical Perspectives*. New York: Springer, 1–22.

Rakotsoane, F. C. L. (2005). The Impact of African Holistic Cosmology on Land Issues: A Southern African Case. *Journal for the Study of Religion*. 18 (1): 5–15.

Ramose, M. B. (1999). *African Philosophy Through Ubuntu*. Harare: Mond Books.

Ramose, M. B. (2003). Ubuntu Philosophy. In, P.H. Coetzee and A.P.J. Roux (Eds.) *The African Philosophy Reader: A Text With Readings*. New York: Routledge, 270–280.

Samkange, S. and Samkange, T. M. (1980). *Hunhuism or Ubuntuism: A Zimbabwean Indigenous Political Philosophy*. Harare: Graham Publishing.

Rathbone, M. and Verhoef, A. H. (2021). Towards a Critical Ethic of Land in the Southern African Context. In, Erasmus Masitera (Ed.) *Philosophical Perspectives on Land Reform in Southern Africa*. Cham: Palgrave Macmillan, 267–284.

Rolston, H. (2000). The Land Ethic at the Turn of the Millennium. *Biodiversity and Conservation*. 9 (2000): 1045–1058.

Tafira, K. (2015). Why Land Evokes Such Deep Emotions in Africa. *The Conversation*. Retrieved from: https://theconversation.com/why-land-evokes-such-deep-emotions-in-africa-42125. Accessed (20 April 2021).

Verhoef, A. H. and Rathbone, M. (2015). A Theologically Informed Ontology of Land in the Context of South African Land Redistribution. *Journal of Southern Africa*. 152 (July): 156–170.

White, L. (Jr) (1967). The Historical Roots of Our Ecological Crisis. *Science*. 155 (3767): 1203–1207.

4 African Relational Environmental Justice

4.1 Introduction

The question of what an African relational conception of environmental justice ought to look like and how it could be different from other non-African and non-relational conceptions of environmental justice is a crucial one. However, this question has not been adequately addressed, despite the existence of a substantial body of literature on African relational ethics. (see, for example, Murove, 2004, 2009, 2014; Metz, 2012, 2019; Mweshi, 2019). One would therefore expect to see how relational views of environmental ethics might provide a basis upon which to conceive an African relational conception of environmental justice. Most influential non-African appeals to environmental justice, especially those in Western philosophy, lean on aspects such as rationality, sentience and welfarist views as grounds on which to conceptualise the fair distributive patterns of environmental justice. In this chapter, I pay attention to the nature of relationships among various beings in the African ontological realm. I then argue that the nature of relationships among these beings determines the nature of African environmental justice. I call it African relational environmental justice because it is based on relationships among various beings.

In general terms, the notion of African relational ethics has been examined by quite a number of philosophers from different angles. These include African communitarianism (Menkiti, 1984, 2004; Gyekye, 2010), the perspective of ubuntu (Samkange and Samkange, 1980; Ramose, 1999) or simply the conceptualisation of relationships (Murove, 2004, 2014; Metz, 2012, 2019). However, in spite of these different approaches, it still remains to be demonstrated how the normative transition is made from human relationships to construct other relationships involving human beings and non-human beings and how conceptions of environmental justice might be gleaned from such relationships. Mweshi has done this by viewing the "appreciation of interconnectedness and interdependencies" (Mweshi, 2019: 194) among beings as grounds for environmental ethics and justice. His account is very important, but it does not go far enough to show the various moral obligations and duties that each of the interrelated and interconnected beings have. I therefore

DOI: 10.4324/9781003176718-5

analyse the notion of relatedness in order to flesh it out as a transcendental and holistic African meta-ethic that goes beyond mere human relationships to include the duties associated with environmental justice that humans have towards other forms of being.

Contrary to some suggestions that African relational ethics is "essentially anthropocentric, [or] a matter of human relationships" (Horsthemke, 2015: 13) and mostly oriented towards agent-centred partiality (Molefe, 2017: 55), I provide a host of reasons why African relational ethics could be interpreted as promoting impartiality in its environmental justice perspective. I examine how African relational philosophies such as communitarianism, *unhu/ubuntu* and vitalist views could help to strengthen relationships and conceptions of moral status and the idea of duty among various forms of being. In the end, I ultimately demonstrate how such relationships could be conceptualised in order to meaningfully construct relational environmental justice perspectives.

The chapter proceeds as follows: in the first section, I give an outline of the nature of African relational ethics. I provide some of the philosophical grounding for such African relational ethics, which include African communitarianism, *unhu/ubuntu* and vitalism. I consider these to be key to understanding the nature of African relational ethics. In the next three sections, I then proceed to propose how relationships among different beings actually form the basis for understanding and ensuring environmental justice for different forms of being. I look specifically at three different kinds of relationships in these three sections, namely, those: (a) between human beings and other human beings, (b) human beings and non-human beings and (c) human beings and non-animate beings. For each of these relationships, I consider the ethical grounds on which to appeal for direct duties and ultimately, moral status, such that environmental justice can be equitably distributed among these related beings.

4.2 A Look at African Relational Ethics

In this section, I provide an outline of what is meant by African relational ethics. I do so because an understanding of African relational ethics is important to the conception of African relational environmental justice that I will examine in the following sections. This will perhaps contribute to the quest for African environmental justice based on conception of harmonious existence among human beings and other non-human beings. According to Mweshi, "understanding justice within the framework of the African conception of harmonious relations allows us to broaden the concept of environmental justice and has a lot to contribute to our African perspective on environmental justice" (Mweshi, 2019: 192). I therefore consider some of the most influential ethical theories in African philosophy such as the African communitarian philosophy, *unhu/ubuntu* philosophy and the African vitalist view, which are key to an understanding of African relational ethics, which in turn forms a plausible basis for understanding African relational justice.

Firstly, one of the most important African perspectives that needs to be explored in African relational ethics is the communitarian view of existence. The African communitarian view of existence is one of the most defining and distinguishing features of African social structures (Gyekye, 2010: 102). It permeates various other dimensions of African relational existence. The African communitarian perspective, as a relational ethical perspective was mainly used as a socialist ideology by African nationalist thinkers such as Nkrumah, Nyerere and Senghor in their quest to resist colonialism as a united African people and to attain independence. As Gyekye notes:

> the advocates of the ideology of African socialism, such as Nkrumah, Senghor and Nyerere, in their anxiety to find anchorage for their ideological choice in the traditional African ideas about society, argued that socialism was foreshadowed in the African traditional idea and practice of communalism (communitarianism)
>
> (Gyekye, 2010: 104).

In African academia, the African communitarian perspective was popularised by Mbiti (1969), Menkiti (1984, 2004) and Gyekye (2007, 2010). In this perspective, Menkiti (1984, 2004) and Gyekye (2010)'s conceptions of person and community, especially their insistence that the individual person is characteristically communitarian, might be taken as generally speaking to the kind of relationships that human beings ought to have in society (Gyekye, 2010: 103). Menkiti, for example, appeals to John Mbiti's (1969: 141) famous dictum that "I am because we are, and since we are, therefore I am", inviting us to understand it as pointing to the affirmation of communal, relational and humane existence based on what it means to exist and relate with others in community. For Menkiti,

> Its [Mbiti's dictum] sense is not that of a person speaking on behalf of, or in reference to, another, but rather of an individual, who recognizes the sources of his or her own humanity, and so realizes, with internal assurance, that in the absence of others, no grounds exist for a claim regarding the individual's own standing as a person. The notion at work here is the notion of an extended self
>
> (Menkiti, 2004: 324).

From this view, Menkiti is convinced that group solidarity is the key or defining feature of African traditional life based on what he sees as the notion of an extended self (Menkiti, 2004: 324). According to this position, the self is extended by virtue of one's inability to exist without others. Hence, it would be proper to say that the individual is essentially relational. This communitarian and largely relational understanding of persons is also later accepted by Gyekye (2007, 2010: 103) and Kalumba (2020), although they, unlike Menkiti, subscribe to a somewhat moderate view of communitarianism.

In the above communitarian view of the person, Menkiti is not so explicit about the relational dimension. However, what matters here is the search for what could be the basis for such solidarity and communitarian existence (which I take as implying relationships). It is also important to see in what sense the individual could be construed as an extended self. In other words, this view saliently refers to the way in which the individual is related to other persons who are similar or different from him/herself in various ways. Menkiti tries to connect the relational character of African ethics to communitarian existence. His position might be interpreted to be anthropocentric because of its over-emphasis on human language as the unique feature of humanity. Nevertheless, it remains important in demonstrating the relational nature of human beings, although it simultaneously shows Menkiti's particularly anthropocentric view of environmental ethics. This view is especially implicit in his conception of animal ethics where he sees animals as having no rights at all (Menkiti, 1984: 177). In this section, I am not interested in his arguments against animal rights, but rather with his conception of communitarianism which I think still accommodates non-human beings. He argues that:

> I have in mind here the lucid example of the human navel and the way it points us to umbilical linkage to biological generations going before. And I have in mind, also, the fact that human language, which is a bio-logically anchored fact, points us, one and all, everywhere in the world, to a mental commonwealth with others – others whose life histories encompass past, present, and future. In both of these examples, biology intimates a message, not of beingness alone, but of beingness together. And to the extent that morality demands a point of view best described as one of beingness-with-others, to that extent does deep biology link up nicely with the direction of movement of the moral order
>
> (Menkiti, 2004: 324).

In the above exposé, Menkiti opines that relatedness among human beings is mainly informed by the possession of a characteristic that is common to all beings, which is language. This is essentially an anthropocentric position, which is largely informed by a limited conception of persons based on sen-tience and language. However, this view remains important in the mapping out of African relational ethics because it attempt to understand what unites human beings with other beings. Gyekye weighs in with the argument that some communal aspects like shared communal values, goods and ends form the basis on which relationships ought to be safeguarded in an African com-munitarian context. He notes that with "…its emphasis on communal values, collective good and shared ends, communitarianism invariably conceives the person as *wholly* constituted by social relationships…" (Gyekye, 2010: 103). Implicitly, therefore, the inseparability of persons from others means that persons are essentially beings who are always in ethical relationships and have connections with other human and non-human beings.

Over and above African communitarian perspectives, there is also an African philosophy based on *unhu/ubuntu*. According to such a view, the quality of being human, or what it means to be human (humanness) is also largely influential towards a conception of humaneness. This ethical understanding of human existence is indeed important to African relational ethics and quite central to a conception of relational environmental justice. Samkange and Samkange are credited for writing the first piece of work on the African philosophy of *unhu/ubuntu* in 1980. They argue that *unhu/ubuntu* as a relational African philosophy ought to be taken as a useful tool in African social and political philosophy (Samkange and Samkange, 1980). Samkange and Samkange (1980: 38–9) invite us to look closely at the famous Shona dictum (which is also expressed in other African languages such as Ndebele, Zulu and Xhosa) that: *munhu, munhu navanhu* (Shona)/*umuntu ngumuntu ngabantu* (Ndebele/Zulu). This expression articulates an African relational conception of existence among human beings because it means that, human beings are communitarian by virtue of existing with others, and that they are essentially related to each other by virtue of existing together. Such a perspective is not unlike Menkiti and Gyekye's positions examined above. For Samkange and Samkange (1980), moreover, not only are human beings related and interconnected, but that there must be some close connections between relatedness and the duty towards such related fellow beings. This is implied in the examples that they provide. They insist that relationships, either through family ties, totems or mere social interactions between people who do not know each other, ought to be respected and safeguarded by values of *unhu/ubuntu* (Samkange and Samkange, 1980: 38). By way of example, these scholars note that, it is not expected that a stranger in desperate need of assistance such as directions, a car-push, drinking water or food, etc. – would be charged a fee for this. This is because as human beings, we are related by virtue of being human beings. At the same time, harging for providing help goes against the values enshrined in *unhu/ubuntu,* which bind humans as related persons.

Ramose (1999) presents one of the most lucid exposé of the philosophy of *unhu/ubuntu* as a complete African relational philosophy that should be read from various angles. He argues that it should be able to give humans guidance in areas such as metaphysics, epistemology and ethics. According to Ramose, "the African tree of knowledge stems from *unhu/ubuntu* with which it is connected indivisibly. *Ubuntu* then is the wellspring flowing with African ontology and epistemology" (Ramose, 1999: 49). Based on this view, *ubuntu* might therefore be taken as the basis for human relationships and even conceptions of duties for justice towards each other and towards other non-human beings. *Unhu/Ubuntu* is therefore a relational ethics, which will be useful to our understanding of environmental justice in African relational ethics because of its accommodation of duties not only towards fellow persons, but also to other beings that humans may not be directly related to because environmental justice demands conceptions of justice beyond human communities. American social contract theorist, John Rawls gives a similar relational perspective of environmental justice as he envisages a perspective of intergenerational justice that also takes into consideration the need for ethical concern and fairness, which is based on the *just-savings* principle (Rawls, 1971: 118–119). This means that the community is incomplete if it does not take into account relationships among different generations, and their import for environmental justice. I will examine this perspective of intergenerational environmental justice in Chapter 6.

In addition to the African communitarian perspectives discussed above and the theories of *unhu/ubuntu*, there are other perspectives that neither lean exclusively towards communitarian nor to *unhu/ubuntu* ethics. Such perspectives borrow from both philosophies as they further interpret African relational ethics from these African philosophies directly. I will simply refer to such conceptions as African relational views of existence. According to scholars who have written on these relational views (see, Murove, 2004, 2009, 2014; Metz, 2012, 2019; Mweshi, 2019), the granting of moral status, rights and duties of environmental justice to beings is based on the sort of connections and relationships they have with each other. For Metz, such African relational attitudes are mostly "grounded in salient sub-Saharan moral views, roughly according to which the greater a being's capacity to be part of a communal relationship with us, the greater its moral status" (Metz, 2012: 387). Metz's position is very important to a conception of African relational ethics, and I will interpret it later in terms of its import to environmental justice. Metz's view is a relational theory that is based on a conception of moral status that is largely informed by either direct or indirect duties to other beings. However, his view could be taken as attributing greater moral status to beings on the basis of their degree of relationships with human beings. Hence, Metz might be seen as accepting anthropocentric thinking. According to Molefe, such a relational view is based on partiality (Molefe, 2017: 53). Notwithstanding such an objection, the attempt to understand the moral relationship between human beings and non-human beings on the basis of what sort of relationships exist between such beings is an important conception of relational ethics in African philosophy.

A similar conception of African relationalism is also presented by Murove (2014) and more recently by Mweshi (2019). Although Murove has examined relational ethics by limiting it to the Shona people of Zimbabwe, his presentation of the notion of *ukama* (relatedness) aptly sums up all these different conceptions of African relational ethics. Murove uses *ukama* and *unhu/ubuntu* to capture the the ethical implications that human relationships have for other spheres of life in general based on such relatedness and *unhu/ubuntu* (humanness), from which humaneness (or consideration for others) is derived. According to Murove, "while the Shona word *ukama* means relatedness, *ubuntu* implies that humaneness is derived from our relatedness with others, not only to those currently living, but also through the generations, past and present" (Murove, 2004: 196). Murove's view could be taken to mean that *unhu/ubuntu* (humanness) is derived from conceptions of relationships (ukama). A detailed discussion of that dimension (i.e., that ubuntu is derived from relatedness) would be necessary and interesting although it cannot be pursued here. Mweshi, on the other hand, is concerned not with *ubuntu* as such, but with the view that "our uniqueness as human beings is overlaid by the interconnectedness and interrelatedness that characterise the whole of reality" (Mweshi, 2019: 194). However, what is important, at this juncture, is to emphasise that, African relational ethics based on communitarian relationships and *unhu/ubuntu* form a chain of ethical connectivity. This ethical connectivity binds human beings with other human beings and non-human beings including animate and non-animate reality, past, present and future generations, a view that I explore later on.

In addition to these perspectives, from African communitarianism and *unhu/ubuntu*, one often neglected position that anchors African relational ethics is the vitalist view of existence (Tempels, 1959; Metz, 2014; Molefe, 2018). The vitalist perspective is based on authenticating relationships between almost all beings that are believed to exist in African ontology, including spiritual beings. For Molefe, "one can safely observe that a dominant interpretation of African tradition espouses an ontology that goes beyond the empirical" (Molefe, 2018: 23). Notwithstanding its metaphysical, supernatural and sometimes superstitious premises, for which it is often criticised, the African vitalist view emphasises on harmonious relationships in the hierarchy of existence among all beings. This somewhat holist attitude (Molefe, 2018: 23) is aptly captured by Ikuenobe as follows:

> Traditional African views of ontology can be understood in terms of their view of cosmology. Reality is seen as a composite, unity and harmony of natural forces. Reality is a holistic community of mutually reinforcing natural life forces consisting of human communities (families, villages, nations, and humanity), spirits, gods, deities, stones, sand, mountains, rivers, plants, and animals. Everything in reality has a vital force or energy such that the harmonious interactions among them strengthen reality
>
> (Ikuenobe, 2014: 2).

From the vitalist position articulated above, a complex network of relationships exists amongst human beings and with other non-human beings. These beings include the spiritual beings, nature, the past, present and future generations. All these beings are vitalist in that they will always have an effect on conceptions of African environmental justice because of the duties and obligations which one being has towards the other. This explains why it must be taken as a holist view of environmental ethics.

In this section, I consider communitarianism, *unhu/ubuntu* and vitalism as constituting important ingredients of African relational philosophy. This is because they are all essentially views of existence that emphasises relational, harmonious and humane existence between human beings and the surrounding world. However, in the quest for African environmental justice conceptions based on African relational ethics, there is need to be cautious of some about the objections raised against it. Some scholars, for example, think that conceptions of animal ethics and environmental justice based on relationalism are inherently anthropocentric, especially in the way such relationships are construed between human beings and nonhuman animals through aspects such as taboos and totems (Horsthemke, 2015: 98, 242). Likewise, others feel that African relational ethics might also be seen as being agent-centred and justifying partiality, thereby failing to value non-human animals for their own sake (Molefe, 2017: 53). These objections are indeed reasonable and cannot be ignored because they clearly highlight some grey areas relating to some of the anthropocentric elements in African relational areas. It would therefore not be possible for African relational ethical theory to be watertight to the extent that it is purely non-anthropocentric. However, despite these important objections, I show that African relational ethics is a reasonable foundation on which to appeal for what could be termed African relational environmental justice, a perspective that I examine in the next section.

4.3 Distributing Environmental Justice through Relationships

In African ontology, the major ethical principles enshrined in African relational ethics can contribute to the conception of how to distribute environmental justice among various related beings ranging from human beings to non-human beings as I will show in the following sections. I will consider how the questions of environmental justice could be meaningfully constructed using the various conceptions of relationships. By environmental justice questions, I am referring to the fundamental issues of rights, justice, fairness, equity and possibly involvement in the distributive patterns of environmental benefits and burdens. This conception of environmental justice is in sync with the American Environmental Protection Agency's understanding of environmental justice, which I examined in the second chapter. However, I seek to propose a broader conception that is not limited to the human community alone. According to the American Environmental Protection Agency, "environmental justice is the meaningful involvement of all people regardless of race, colour, national origin or income, with respect to the development, implementation and enforcement of environmental laws, regulations, and policies (Environmental Protection Agency, 2008). A similar view is also proffered by Bullard, who is also concerned with disparities in the patterns of waste disposal, particularly that waste is disposed closer to black communities in twentieth century America (Bullard, 1990). These two definitions of environmental justice were developed within a non-African context, being focused mostly on the American context. However, they are very useful to understanding environmental justice conceptions in Africa, although I have two points of contention with both the Environmental Protection Agency (2008) and Bullard (1990)'s understanding of environmental justice. My conception of African environmental justice goes beyond the two views which are largely anthropocentric.

Firstly, the EPA's conception of environmental justice is essentially anthropocentric. Its anthropocentric understanding of environmental justice is implicit in limiting environmental justice to concern with the meaningful involvement of *all people* in the distributive patterns of environmental benefits and burdens. The anthropocentric slant of this understanding is evident in the use of "all people". It is silent about the import of other non-human beings for the discourse of environmental justice. This, is despite the fact that human beings share fundamental relationships that also have ethical implications for other beings with which they relate. Although it is not possible to actively involve non-human beings in environmental policy-making and implementation, to which the EPA definition is mostly limited, environmental justice should at least, also take into account the interests, welfare and well-being of such beings.

Secondly, I do not intend to limit questions of environmental justice to issues of environmental burdens as if they are beginning to be experienced presently. I will argue that a conception of environmental justice has always been inherent in African ontology, particularly in African relational ethics where environmental benefits such as the land, natural resources, wild animals, green spaces and minerals on the one hand, and environmental burdens such as pollution, natural resource depletion or extinction and poverty on the other, were equitably distributed. Ultimately, I will show that African relational ethics meaningfully involves almost all beings in such distributive patterns of environmental justice.

In the following three sections, I will first look at the relationships of human beings with their fellow human beings in terms of how they could meaningfully influence the distributive patterns of environmental justice, taking into account the duties and obligations of justice that human beings have towards each other. I will then proceed to consider the duties and obligations that human beings have towards non-human animals in terms of how they contribute, indirectly, to a conception of environmental justice based on the human-animal relationships. At the end, I expose the other indirect duties to environmental justice that human beings have by virtue of their relationships with aspects of non-human reality such as water sources and the air.

4.4   Human Relationships and Duties to Fellow Human Beings

In the first section in this chapter, I argued that human beings are connected to each other by various relationships. These relationships range from familial relationships to extra-familial relationships, which include other human beings and non-human beings. In this section, I seek to show the basis for such a relational-ethical perspective. I do so by attempting to establish where the duties towards other human beings might come from in thinking about environmental justice issues. I seek to demonstrate how this relational ethics of duty is significantly oriented towards environmental justice. I show how, according to the African relational view, the duties that individuals have towards fellow human beings could be read as the basis for environmental justice at large.

Firstly, human relationships in African relational ethics are mainly anchored on, and strengthened by the ethics of duty. Because the African community is largely communitarian, it essentially holds that the individual does not exist in a vacuum as a lone individual (Menkiti, 1984, 2004; Gyekye, 2007, 2010), but in relationships with others, it naturally follows that individuals have certain duties to other human beings or community at large, since communitarianism is not optional (Gyekye, 2010). This perspective is also shared by Kalumba who argues that relationships to the community are necessary and involuntary (Kalumba, 2020: 138). As opposed to what Kalumba sees as an "Individual Moral Primacy thesis", which is characteristic of individualistic theories of social and political thinking in non-African traditions, African communitarianism upholds the "Communal Moral primacy thesis, which prioritises the community and its claims over the individual and his or her claims" (Kalumba, 2020: 138). This means that the duties that the individual person has towards fellow human beings are based on the fact that communitarian life is not optional for the individual in the first place, at least from Gyekye's perspective because the person is not complete without the community. This view is captured in the following assertion:

> The communitarian conception of the person has some implications: it implies (i) that the human person does not voluntarily choose to enter into human community, that is, that community life is not optional for any individual person; (ii) that the human person is at once a cultural being; (iii) that the human person cannot – perhaps must not – live in isolation from other persons; (iv) that the human person is naturally oriented toward other persons and must have relationships with them; (v) that social relationships are not contingent but necessary; and (vi) that, following from (iv) and (v), the person is constituted, but only partly (see below), by the social relationships in which he necessarily finds himself
> (Gyekye, 2010: 105).

From Gyekye's view, also shared by Kalumba above, the individual is naturally duty-bound towards the community in which s/he naturally finds him/herself. In essence, "the fundamentally relational character of the person and the interdependence of human individuals arising out of their natural sociality are, thus, clear" (Gyekye, 2010: 105). This brings in the dimension that even on issues to do with environmental justice, that the relational character of persons, taken together with their interdependence will inform a positive conception of environmental justice.

From the above communitarian perspective, African relational ethics and its conception of environmental justice might not be necessarily individualistic. Nor are they based on traits within the individual such as rationality, as articulated in the Kantian view. Because of the communitarian underpinnings, it places strong emphasis on the duties that individuals have towards fellow human beings. African communitarian relational ethics can therefore be taken as an ethics of duty towards the community or fellow beings. This view could then be taken as the basis for understanding environmental justice based on such ethics of duty. It must be emphasised that within the discourse of African environmental justice in general, human beings, irrespective of their disposition, have certain duties of justice towards each other (Menkiti, 1984: 177) in the distributive patterns of environmental benefits and burdens. However, while Menkiti's view that human beings are owed such duties of justice is philosophically plausible, I do not accept his radical anthropocentrism that stems from such an idea, since opines that only human beings deserve such duties of justice and therefore essentially limits moral status to human beings alone, in keeping with Kantian ethics.

African relational ethics gives individuals a sort of prima facie duties to the communitarian relationships so that individuals can conform to what is right and good for the community. The ethics of prima facie duties is essentially based on doing things that "matter morally; and make a difference to what we should do and to whether we acted rightly in what we chose to do" (Dancy, 1991: 221). Human beings, for example, can choose to help people facing poverty in society, and their actions should be right to that extent because they have agreed that they will help the poor. It is actually right to choose to help those that are poor. It is a prima facie obligation to help such people according to the duty of improving the welfare of others (the prima facie duty of beneficence) (Dancy, 1991: 221). In African relational ethics, similar duties and obligations come from the loose understanding of justice as "what is right, and what human beings have a duty to do" (Ikuenobe, 2014: 10). From this perspective of African relational ethics, following the above conception of duty, the positive or negative benefits or impact from the environment ought to be considered in terms of their impact on fellow human and non-human beings. Because of the relational structure of African society, and its emphasis on the duty for beneficence based on such relationships, consideration ought to be given as to how not to distribute goods in ways that unfairly discriminate others. Accordingly, human beings ought to relate with the environment is such a way that they can continue to live well with other human and non-human beings and the environment at large. Although the origin of such duties may not be so clear, relationships among beings, I argue, should be taken as the basis for rights and duties for the equitable distribution of environmental goods. In African ontology, relationships are a good ground for environmental justice. This is because relationships are

not only limited to human beings, therefore allowing for a broader conceptualisation of environmental justice. Relationships in African ethics could be
meaningfully interpreted by focusing on human beings in as much as they
are related to each other, to non-human animals, to non-human reality and
to past, present and future generations.

The variety of relationships, i.e., among humans and other non-human
beings might be read as potentially contributing to a conception of the altru-
istic duties towards environmental justice because all beings are thought to
be related. That should be good grounds for duties towards the attainment of
environmental fairness and equity among human beings. Although Kelbessa
is not very explicit about African relational ethics as such, he makes the following important observation:

> In many African societies, the members of the clan include the unborn,
> those living in the world of ordinary sense experience, and those living
> in the post-mortem world of the ancestors. According to the African
> worldview, currently living human and nonhuman beings, the living
> dead, the yet unborn, and the natural world are interconnected. For many
> African communities, it would be wrong to over-consume resources and
> leave future generations with fewer means of survival
>
> (Kelbessa, 2015: 50).

Thus, based on Kelbessa's argument, human beings are duty-bound towards
environmental justice to fellow human beings, non-human animals, non-human
reality and future generations. This view of relational environmental justice
is premised on the relationships, and ultimately the duties and obligations
that individuals have, not towards fellow individuals to which they are indi-
vidually related per se, but also towards other communities of which they
are an inseparable part, and towards which they ought to have duties. Such a
view makes sense, considering what Gyekye sees as the "natural membership
to the community, which of course cannot rob the individual of his/her dignity or worth, a fundamental inalienable attribute he/she possesses as a person" (Gyekye, 2010: 115). From this view, notwithstanding the autonomy of
the individual, which s/he cannot be denied for the mre fact of being a part
of the community, the obligations that individuals have towards the community still arise from that aspect of communing, making it possible for individuals to be humane towards others, or fellow members of the community.

Implicit in African relational ethics is also the humanist philosophy of
unhu/ubuntu which I also treat as the source of ethical values for members
of an African relational community. Indeed, it is not possible to have a suc-
cessful discussion of African relational ethics without making reference to
unhu/ubuntu because they are closely knit together. African relational environmental justice is also concretised by the ethics of *unhu/buntu* because that
is what defines human beings. As a relational ethics, *unhu/ubuntu* emphasises
that human beings have certain obligations to fellow beings on the basis of

what a human being is understood to be. Such a philosophy is thought to give human beings certain duties and obligations towards each other, and ultimately for environmental justice because of its humane underpinnings. According to Kelbessa, "ubuntu philosophy recognises entitlements and obligations towards others" (Kelbessa, 2015: 62). Thus, *unhu/ubuntu* becomes a useful framework on which to appeal for environmental justice because it speaks to and respects the claims and entitlements of different individuals, including non-human beings, as I will demonstrate.

Unhu/Ubuntu remains one of the most fundamental source or foundation of duty in African relational ethics of duty towards fellow human beings and other non-human beings as well. The notion of duty that the individual has towards fellow human beings in *unhu/ubuntu* can be read from Ramose' s etymological definition of the term *ubuntu* as derived from two conjunctions – *ubu* implying the notion of *be-ing* or existence in general and – *ntu* signifying the nodal point at which *beingness* achieves fullness or concrete form (Ramose, 1999: 49–50). The attainment of *beingness* or fullness in this regard ought to be closely read with reference to the duties that are associated with that person or *being*. Chivaura gives an almost similar etymological analysis to that given by Ramose, but that his view aptly captures the aspect of duty which the individual carries through *ubuntu*. For him, "the -*nhu* in *hu-nhu* or -*ntu* in *ubu-ntu* refers to one's physical existence as a *thing* with no values attached. *Hu*- and *ubu*- indicate values. People who lack *hu*- or *ubu*- attached to them are mere –*nhus/-ntus* or *things. Havana unhu,* in Shona: they lack human content" (Chivaura, 2006: 232). Both Ramose and Chivaura's etymological interpretations of *unhu/ubuntu* point to the view that *hunhu/ubuntu* is a humanistic ethic. Such a conception sees moral humaneness or ethical existence and consideration for other persons to be based on what it means to be a human person first, and to have certain expectations and duties associated with being a human person.

At a more practical level, the ethics of duty according to which individuals ought to have duties towards fellow human beings is also enshrined in the African Charter on Human and People's Rights (ACHPR) that was adopted in Banjul in 1981 by the Organisation of African Unity (OAU, 1981), now the African Union, (AU). This Charter was mainly informed by African relational ethics, and seeks to contextualise and moderate the human rights discourse in Africa so that it is not understood from an exclusively individualistic perspective. The African Charter on Human and People's Rights not only taps from African relational ethics, but that it places emphasis on the duties that individual persons ought to have towards other human persons. Articles 27, 28 and 29 of the ACHPR, for example, explicitly capture these duties as follows:

> Article 27 (1): Every individual shall have duties towards his family and society, the state and other legally recognised communities and the international community.

Article 21 (2): The rights and freedoms of each individual shall be exercised with due regard to the rights of others, collective society, morality and common interest.

Article 28: Every individual shall have the duty to respect and consider his [/her] fellow human beings without discrimination, and to maintain relations aimed at promoting, safeguarding and reinforcing mutual respect and tolerance

(ACHPR, 1981).

From these articles, what comes out clearly is the emphasis on aspects such as 'community', 'rights of others', 'common interests', 'respect', 'mutual respect' and 'tolerance'. These are central to one's duties towards other human beings and inform one's conception of these duties. Such a view of duty might be a plausible basis for the equitable distribution of environmental justice, since it takes into account the rights of others, common interests, the need for mutual respect and tolerance in the use of the environment. As articulated in article 27 (1), by virtue of the individual person having duties to family and society, it means that his/her relationships with the environment and use of nature's resources ought to take into account the duties that s/he owes to family and society. This means that an individual's moral actions are largely guided by the extent to which such actions impact on other human beings towards whom s/he has duties. This is a reasonable view of environmental justice for a duty-based view of environmental justice.

I have only considered here, the relationships between and among human beings and the duties and obligations that they might have towards each other. In order for any view of environmental ethics and justice to be non-anthropocentric, however, it must go beyond the human-to-human relationships. In the following section, I therefore focus on the relationships between human beings and non-human animals. I seek to establish the duties and obligations that human beings might have towards non-human animals and how such a conception might contribute to environmental justice.

4.5 Human–Non-Human Animal Relationships and Duties to Non-Human Animals

In the previous section, I examined the way human beings have direct duties towards each other on the basis of their relationships. In this section, I seek to show what implications the nature of human–non-human animal relationships in African relational ethics have for conceptions of environmental justice. I start by responding to some of the objections that might be raised against African relational environmental ethics based on relationships between humans and non-human animals. I then proceed to analyse the nature of duties between human–non-human animal relationships, the basis for such duties and how such conceptions of duty might be meaningfully understood to influence conceptions of environmental justice between human beings and non-human animal relationships.

So far, a number of philosophers have examined African relational ethics by focusing on the relationships between human beings and non-human animals in terms of the import of those relationships for environmental ethics (Murove, 2004, 2009, 2014; Metz, 2012, 2019; Horsthemke, 2015; Chemhuru, 2016, 2019; Molefe, 2020). However, these accounts appear to be silent on the conception of environmental justice per se. Most of these accounts are merely concerned with the environmental ethical import of such a relationship without interrogating further the kind of duties that human beings ought to have towards non-human animals, and how such duties have a bearing on fellow human beings in the distributive patterns of environmental equity. In *Animals and African Ethics*, Kai Horsthemke (2015) puts forward a conception of African environmental justice based on the relationships between human beings and animals. Ultimately, just like Metz (2012, 2019), Horsthemke (2015) seems to be taking African relational ethics and its conception of environmental justice into some form of anthropocentrism. However, he later departs from his earlier position and concedes that "there exist several resources in African philosophical thinking for deriving a non-anthropocentric and non-speciesism ethical orientation" (Horsthemke, 2017: 183). This latter position is one that I will defend as I look at African relational environmental justice as non-anthropocentric.

Another objection that is often raised against African relational ethics and its view of environmental justice is that it is agent-centred or oriented towards agent partiality. This is because relationships between human beings and non-human animals are thought to revolve around the human being as the main moral agent (Molefe, 2017: 55). The other reason is that by nature, "human beings are at the core of these communities …[such that] it is difficult not to interpret this as 'human-centred'" (Horsthemke, 2015, 2017). However, reasonable this view might be, I maintain that African relational ethics is not oriented towards agent-related partiality. Indeed, African relational ethics through what Murove (2004, 2009, 2014) sees as the ethics of *ukama,* brings together relationships between human beings and non-human beings. According to Murove, *ukama* "is also based on the totemic system where by a person sees himself or herself as related to natural species, thereby instilling a sense of belonging to the wider environment, past as well as future" (Murove, 2014: 44). This is an environmental ethical perspective that has been examined by, for example, Chemhuru and Masaka (2010) and Mangena (2013). These philosophers concur that African relational ethics is not only relationally oriented towards human beings alone but is also oriented towards other non-human animals. The value of such a view to environmental justice is therefore implicit. According to this view of relational environmental justice, certain animal species are honoured and respected as totems or totem animals because they are related to human beings. Masaka concurs when he avers that "totems are aimed at creating harmonious and respectful co-existence between human beings and non-human animals and the environment" (Masaka, 2019: 35). Consequently, by belonging to different totems based on certain groups of animals, human beings are taught that they at least have relationships with these animals. That perspective would teach human beings to accord non-human animals some sort of direct moral status and avoid cruelty towards them. Notwithstanding accusations of propagating a somewhat human-centred ethics based on constructing indirect relationships between human beings and non-human animals, these views remain central to an understanding of human relationships in terms of their import to conceptions of environmental justice, as I demonstrate below.

The idea of moral status remains at the core of African relational ethics and environmental justice conceptions relating to the human–non-human animal relationships. As opposed to environmental justice based on human interests and justice for humans alone, moral status, particularly that of non-human animals, ought to be the basis upon which African conceptions of environmental justice are premised. According to Metz, "the concept of moral status is the idea of something being the object of a 'direct' duty i.e., owed a duty in its own right, or is the idea of something that can be wronged" (Metz, 2019: 11). According to this view, the idea of animals having moral status is mainly informed by their status as objects of direct duty, and it is clear that animals are the objects of direct duty because they can be wronged through various actions that human beings may inflict to them in their relationships with them. Of course, Metz reminds us that "both animals and human beings have moral status of the same kind, [but] different in degree [such that] even a severely mentally incapacitated human being has greater moral status than an animal with identical internal properties; and a new born infant has a greater moral status than a mid-to-late stage foetus" (Metz, 2019: 9). This view merely emphasises the relational properties of an African theory of moral status and the way moral status varies from one being to the other. This explains why in African ontology, even if human beings have direct duties towards non-human animals, they continue to use animals for various purposes such as the creation of animal products and agriculture through animal labour, etc. However, this view should not be taken to mean that human beings do not have direct duties to such animals and that animals do not have moral status, or that African relational ethics does not respect some intrinsic properties in animals. This is where I depart from Metz's view. He holds that following an African modal-relational account of moral status, non-human animals cannot be granted moral status on the basis of their intrinsic properties, but on the ability of a being to participate in social relationships (Metz, 2019: 12). I do not agree with Metz in this regard because I think that the fact that human beings having direct duties to animals means there must be some intrinsic properties in animals that give human beings such duties. Such intrinsic properties could include the fact that, non-human animals are subjects of a life (Regan, 1987: 186), they have sentience (Singer, 1975) and live independently of human beings. Prior to these views in Western philosophy,

Jeremy Bentham and John Stuart Mill had tried to at least go beyond the speciesist and agent-related moral view (Molefe, 2017: 55) and insist on sentience and welfare, thereby at least taking animate beings into the moral world of relationships on which African relational ethics places emphasis. Within the African context, the notions of sentience and welfare have an extended meaning beyond the human community. This is because "we take the community to be extensive to include every being or non-human nature" (Etieyibo, 2017: 157). This may explain why even the notion of *unhu/ubuntu* is also applied to human-non-human interactions. For example, one who treats non-human animals in a way that is not morally acceptable might be judged as lacking *unhu/ubuntu* because it goes against the direct duties that human beings have towards non-human animals.

The distinction between direct duties and indirect duties in human and non-human relationships is also central to an African relational conception of environmental justice. This conception is shaped by the fact that human beings ought to have direct duties towards non-human animals and is not based on human beings' indirect duties to fellow human beings. Indirect duties refer, for example, to a situation where, if an object is said to have moral value based on indirect duties, "an agent has a moral reason to treat it a certain way but not ultimately because of facts about it" (Metz, 2019: 11). In the work, *Of Duties to Animals and Spirits,* for example, Kant draws the conclusion that human beings do not have direct moral duties to animals (Kant, 1963: 239–242). As a result, according to Kantian ethics, in their relationships with non-human animals, human beings might only have indirect duties towards their fellow human beings and future generations. Such indirect duties oblige human beings not to use non-human animals to the extent that they are at risk of extinction because other human beings and future generations have entitlements to those non-human animals and their products just as current generations do. Such a view is based on the indirect moral obligations to which human beings have towards other human beings and future generations, but it is not informed by the idea of direct duties. This Kantian approach to environmental justice issues is excessively anthropocentric and is seems contrary to the view based on African relational ethics, which lays emphasis on a view of environmental justice based on direct duties towards non-human animals.

As I have just indicated, most accounts of environmental ethics and justice take the duty not to mistreat non-human animals as an indirect duty in which protecting non-human animals is aimed at safeguarding the welfare of human beings. However, African relational ethics is different because human beings have direct duties towards non-human animals since they are a part of the *community* that therefore have duties towards by virtue of their relationships and also the moral status of such beings. Etieyibo articulates this view as follows: "Because it is a community-centred ethic and humans are but just one member of this community, their ethical duty is to promote 'being-spherical egalitarianism' or simply the intrinsic value of every member of the community" (Etieyibo, 2017: 157). Implicit in this view is the egalitarian and duty-oriented view of environmental justice among all beings in the community, including non-human animals. Moreover, the recognition of the aspect of sentience in animals in so far as they are objects that can feel pain and pleasure, means that our actions as human beings can directly harm them. Thus, human beings are directly duty bound not to mistreat non-human animals because they are subject of direct duty and ultimately have moral status, albeit of a lesser degree than that of human beings (Metz, 2019: 9). However, notwithstanding the varying degrees of moral status emphasised in African relational ethics (Metz, 2019: 9), animals, in their own right, can be wronged. So, a conception of how the distributive patterns of environmental justice might be conceived between human beings and non-human animals is not only informed by the interests of human beings alone but also in accordance with the duties owed to non-human animals by human beings. By way of example, the duty to safeguard certain types of animals in Africa, such as the famous *Big Five* (lion, leopard, rhinoceros, elephant and buffalo) as well as other endangered animal species like the gorilla, pangolin, python, penguin and wild dogs is largely informed by the direct moral duties that human beings have towards these non-human animals, and not by the need to protect and preserve them for their fellow human beings, or the degree of their relationship with humans, as implied in Metz (2019). This view concurs with Horsthemke's view that "the way to worry about members of a rare or endangered species is to think of them as individuals whose lives and well-being are under threat" (Horsthemke, 2017: 185). This is a largely non-anthropocentric view of environmental justice based on the nature of relationships between human beings and non-human animals.

Much of the discussion in this section was an attempt to demonstrate the import of African relational ethics to conceptions of environmental justice. I have largely appealed to views from African relational ethics such as those by Murove (2004, 2009, 2014) and Metz (2012, 2019) that have helped me to lace emphasis on duties towards non-human animals. In doing so, I have deliberately avoided some African communitarian views on personhood that I examined in the previous section, especially those of Menkiti (1984, 2004) and Gyekye (2010). This is because I acknowledge that such views are somewhat rejecting the notion that human beings have duties towards non-human animals (Molefe, 2020: 84–5). Nevertheless, their views remain useful to an understanding of relational environmental justice, as I have already demonstrated. In the next section, I consider the nature of relationships between human-non-animate realities, and what duties and obligations might exist in such relationships in so far as environmental justice is concerned.

4.6　Human-non-Human Beings Relationships and Duties to Non-Animate Reality

So far, I have considered environmental justice based on relationships and duties that human beings have towards fellow human beings and towards non-human animals in the previous two sections, respectively. Some might ask, therefore, whether it would be possible to also propose a similar environmental justice framework based on human relationships with non-animate reality. I seek to address the same question in the affirmative and examine what makes these aspects of reality (i.e., inanimate beings) central to the nature of an environmental justice perspective from African relational ethics. By non-animate reality, I am referring to aspects of the environment that include, but are not limited to, collective physical realities such as all *non-animate-living reality* (e.g., all plant species and vegetative life) and all *non-animate-non-living reality* (e.g., water sources, mountains, the land, the atmosphere and the air around them). In considering these two perspectives, I propose an African relational view of environmental justice based on African conceptions of biocentrism and ecocentrism, respectively.

4.6.1 Duties to Non-Animate-Living Reality

Human beings share relationships and the environment with other life forms that are not necessarily human and animate. For example, various plant and vegetative species live lives independently to those of humans and animate beings. The latter beings (i.e., humans and animate) actually depend on the former (i.e., plants and vegetation) for their livelihood in the form of food. Of course, I do not want to take that anthropocentric view as the major premise for my support of the argument for duties towards such beings. I contend that human beings ought to have duties towards other life forms such as plants for non-anthropocentric reasons and reasons relating to the ethical standing of these plant species. One reason for thinking that such living reality warrants direct duties from human beings is that they *share* a relationship of life with them, i.e., both human beings and non-animate-living reality are all objects of a life together. Therefore, such species should be taken as having some sort of intrinsic value (i.e., value beyond their usefulness to human beings (Attfield, 2014; Etieyibo, 2017: 149)) because they all live independent and meaningful lives. This kind of a view of African environmental justice that takes these particular life forms into consideration is largely informed by the relationships between human beings and all life forms. A similar view in Western philosophy is anchored on the biocentric view of environmental ethics. According to such a view, "unlike anthropocentrism which places *humans* at the centre of the universe, biocentrism places the *biosphere* at the centre of the universe" (Etieyibo, 2017: 149). Related to, albeit different from, this biocentric conception of environmental justice is also the ecocentric conception, which I will examine below.

4.6.2 Duties to Non-Animate, Non-Living Reality

In order to ascribe duties to non-animate reality, African relational environmental ethical perspectives animate aspects of non-animate reality such as water sources, mountains, the land, the air around and the atmosphere. The animation of these forms of reality is done so that these non-animate aspects of the Earth can be in some sort of relationships with human beings. By this, I mean that these non-animate realities are considered to be alive, in a sense so that relationships with human beings are easy to understand, even in terms of the import of such relationships for environmental justice. The basis on which aspects of *non-animate* reality could be accorded direct duties is that they are *alive* and also related to human beings. This cannot be taken to be an anthropocentric reason for giving them moral status because the reasons for giving them respect are not oriented towards human beings.

Like non-animate living reality such as plants, trees and vegetation, non-animate and non-living aspects of reality such as mountains, natural water sources (i.e., rivers, dams and seas), the land and the atmosphere are also taken as the abode of other living and non-living beings, including spiritual beings. Magesa, for example, makes reference to ancestors, arguing that they "… are *active beings* who are either disincarnate human persons or powers residing in natural phenomenon such as trees, rocks or lakes" (Magesa, 1997: 35–6). The latter view on spiritual beings might be objected for its metaphysical premises. However, I will not lean on it alone as my defence of relational environmental justice based on relationships between humans and non-living reality. My argument that non-living reality has moral status is largely premised on other relational perspectives based on relationships between human beings and the other beings that exist within the mountains, in water, on the land and in the air around them.

Firstly, it must be emphasised that the duties towards non-animate and non-living realities such as water sources, mountains, the land, the atmosphere and the air around them derive mainly from the view that they are the habitat of all life forms including human and non-human life such as animate life and vegetative life. As a result, a threat to non-animate realities is ultimately a threat to life in general. Human beings ought to have some indirect duties to protect such aspects of reality because that is where life flourishes. Destroying them is as good as destroying life in general. Ultimately, the reason for according them moral status on that basis is not anthropocentric because it is not focused on the relationships of these aspects of reality with human beings alone, but with almost the entire ecosystem. This kind of African relational view of environmental justice for the entire ecosystem might therefore be considered to be ecocentric rather than anthropocentric, because it prioritises the interests of all life forms by taking into account and protecting their habitat. To this end, Etieyibo observes that "unlike anthropocentrism, ecocentrism places the ecosphere at the centre of the universe, it is a *nature-centred ethic* … in an ecocentric ethic, nature has moral worth and consideration because it as intrinsic value" (Etieyibo, 2017: 49–50). In African relational ethics, for example, various myths and taboos around sacred mountains, rivers and water sources are essentially an attempt not only to ensure right relationships between human beings and those aspects of the environment but also to at least give them a degree of respect and autonomy so that human beings do not always see them as existing for their various purposes.

Secondly, the vitalist view is another important albeit controversial perspective based on the metaphysical understanding that all beings are vital forces that are interconnected with each other. It is a useful perspective to African relational environmental justice because of its acceptance in African philosophy and environmental ethics. According to Tempels, "there is no idea among Bantu of *being* divorced from the idea of *force*. Without the element *force*, *being* cannot be conceived" (Tempels, 1959: 151–2). This kind of supernatural, holist and vitalist view is shared by Molefe, who acknowledges that "the African system of reality is characterised by a trilogy of features, namely: supernaturalism, holism and vitalism" (Molefe, 2018: 22). However, the supernaturalist view and the holist view are implied in the vitalist view as well. For this reason, I will focus on the vitalist view below.

Specifically, non-animate reality such as mountains, rivers and the air around are all thought to be *alive* in so far as they have some influence in the life of other living beings such as human beings and non-human animals. For example, it is in these aspects of reality that other vitalist beings such as invisible beings like ancestors are thought to reside. This view was mainly popularised by Tempels (1959), and later taken up by Magesa (1997) and Bujo (2001). According to such a view of existence, among various related beings – ranging from the spiritual beings, human beings, non-human animals and non-animate beings – there are some beings that have more vitality or vital force because of their ability to cause or influence certain phenomenon to happen in the lives of other related beings. Magesa, for example, argues that "ancestors exercise their vitalising influence on the living generations" (Magesa, 1997: 47). This aspect of vitality differs from one being to the other in the same way we have seen in the case of moral status, where despite all beings having moral status, it somewhat differs from one being to the other. Metz elaborately captures this view as follows:

> Life force is traditionally interpreted as a valuable, spiritual energy that inheres in everything, including physical or visible things. Everything in the universe, even an inanimate object such as rock, is thought to be good by virtue of having some degree of life-force, with animate beings having a greater share of it than inanimate ones, human beings having more than plants and animals, ancestors, whose physical bodies have died but who live on in a spiritual realm, having even more than human beings, and God, the source of all life force, having the most
>
> (Metz, 2014: 6962).

Implicit in the above passage is the view that all related beings in the African hierarchy of existence have vitality, or are actually vital forces since they are thought to have some invisible spiritual energy that is considered to be inherent to them all, although it tends to vary in terms of degree of importance from one being to the other. Relationships between living beings and non-living beings must therefore be respected, safeguarded, strengthened and reciprocated all the time in order to attain justice at large, including environmental justice, following the same hierarchy of vital forces. The implications of such a hierarchical view would be that, in matters of environmental justice distribution, despite all reality having vitality or life force, it would not make sense for human beings to prioritise rocks at the expense of, for example, vegetative life, non-human animals and human beings in that order of potency. However, all aspects of reality remain important in the relationships between human beings and non-human reality. Magesa sums up the importance of such relationships by arguing that, "human participation and solidarity, not only with God, the ancestors and other spirits, but also with other elements of creation, are aspects of enhancement of life" (Magesa, 1997: 52). Magesa's view emphasises the relationships that should always subsist between human beings and all forms of being, as I have tried to demonstrate in this chapter.

4.7 Conclusion

In this chapter, I have specifically examined the implications of relationships, ranging from relationships among human beings, including relationships with the various other beings in terms of the implications for environmental justice. In making these considerations, I do not wish to claim that the conception of environmental justice in African relational ethics would be applicable elsewhere, or outside the African context. Nevertheless, such a perspective is worth taking seriously, considering Mweshi's position that "the African emphasis on harmonious relationships offers a plausible way out of the theoretical quagmire that characterises Western environmental ethics" (Mweshi, 2019: 192). Although this perspective is worth pursuing, I have not made such a claim because it is outside the scope of this particular work. However, even if one wished to make such a claim, s/he might still appeal to my relational approach to environmental justice as a framework for such an undertaking elsewhere. The implications of my theory, therefore, would be that since relationships cut across traditions, it might still be possible to conceive environmental justice conceptions in other non-African traditions. Such an enterprise would be worthwhile and could be done using similar relational premises as those that I have used here such as the appeal to communitarian existence, *unhu/ubuntu* and vitality.

References

Attfield, R. (Ed) (2014). *Environmental Ethics: An Overview for the Twenty-First Century*. Cambridge: Polity Press.

Bujo, B. (2001). *Foundations of an African Ethics: Beyond the Universal Claims of Western Morality*. New York: The Crossroad Publishing Company.

Bullard, R. D. (1990). *Dumping in Dixie: Race, Class and Environmental Quality*. USA: Westview Press.

Chemhuru, M. (2016). *The Import of African Ontology for Environmental Ethics*. D. Litt et Phil. (Philosophy) [Unpublished]: University of Johannesburg. Retrieved from: https://ujcontent.uj.ac.za/vital/manager/index?site_name=Research%output (Accessed 1 March 2017).

Chemhuru, M. (Ed.) (2019). *African Environmental Ethics: A Critical Reader*. Cham: Springer Nature.

Chemhuru, M. and Masaka, D. (2010). Taboos as Sources of Shona People's Environmental Ethics. *Journal of Sustainable Development in Africa*. 12 (7): 121–133.

Chivaura, V. G. (2006). *Hunhu/Ubuntu:* A Sustainable Approach to Endogenous Development, Bio-cultural Diversity and Protection of the Environment in Africa. *Presented at the Papers International Conference*. Geneva: 3-5 October, 229–240

Dancy, J. (1991). An Ethics of Prima Facie Duties. In, Peter Singer (Ed.) *A Companion to Ethics*. Malden: Blackwell Publishers, 219–229.

Environmental Protection Agency, (2008). *Environmental Justice*. Retrieved from: https://www.epa.gov/environmentaljustice (Accessed 15 October 2020).

Etieyibo, E. (2017). Anthropocentrism, African Metaphysical Worldview, and Animal Practices: A Reply to Kai Horsthemke. *Journal of Animal Ethics*. 7 (2): 145–162.

Gyekye, K. (2007). *Tradition and Modernity: Philosophical Reflections on the African Experience*. New York: Oxford University Press.

Gyekye, K. (2010). Person and Community in the Akan Thought. In, Kwasi Wiredu and Kwame Gyekye (Eds.) *Person and Community: Ghanaian Philosophical Studies 1*. Washington D.C: The Council for Research in Values and Philosophy, 101–122.

Horsthemke, K. (2015). *Animals and African Ethics*. New York: Palgrave Macmillan.

Horsthemke, K. (2017). Biocentrism, Ecocentrism and African Modal Relationalism: Etieyibo, Metz and Galgut on Animals and African Ethics. *Journal of Animal Ethics*. 7 (2): 183–189.

Ikuenobe, P. A. (2014). Traditional African Environmental Ethics and Colonial Legacy. *International Journal of Philosophy and Theology*. 2 (4): 1–21.

Kalumba, K. M. (2020). A Defence of Kwame Gyekye's Moderate Communitarianism. *Philosophical Papers*. 49 (1): 137–159.

Kant, I. (1963). *Of Duties to Animals and Spirits*. In, *Lectures on Ethics*. Trans: Louis Anfield. New York; Harper and Raw., 239–242.

Kelbessa, W. (2015). Climate Ethics and Policy in Africa. *Thought and Practice: A Journal of the Philosophical Society of Kenya*. 7 (2): 41–84.

Magesa, L. (1997). *African Religion: The Moral Traditions of Abundant Life*. New York: Orbis Books.

Mangena, F. (2013). Discerning Moral Status in the African Environment. *Phronimon*. 14 (2): 25–44.

Masaka, D. (2019). Moral Status of Non-Human Animals from an African Perspective. In, Munamato Chemhuru (Ed.) *African Environmental Ethics: A Critical Reader*. Cham: Springer, 223–236.

Mbiti, J. S. (1969). *African Religions and Philosophy*. London: Heinemann.

Menkiti, I. A. (1984). Person and Community in African Traditional Thought. In, Richard Wright (Ed.) *African Philosophy: An Introduction*. Lanham: University Press of America, 171–181.

Menkiti, I. A. (2004). On the Normative Conception of a Person. In, Kwasi Wiredu (Ed.) *A Companion to African Philosophy*. Malden: Blackwell Publishers, 324–331.

Metz, T. (2012). An African Theory of Moral Status: A Relational Alternative to Individualism and Holism. *Ethical Theory and Practice*. 15 (2012): 387–402

Metz, T. (2014). Vitality, Community, and Human Dignity in Africa. In, Alex C. Michalos (Ed.) *Encyclopaedia of Quality Life and Well-Being Research*. Dordrecht: Springer, 6960–6967.

Metz, T. (2019). An African Theory of Moral Status: A Relational Alternative to Individualism and Holism. In, Munamato Chemhuru (Ed.) *African Environmental Ethics: A Critical Reader*. Cham: Springer, 9–27.

Molefe, M. (2017). Relational Ethics and Partiality: A Critique of Thad Metz's 'Towards an African Moral Theory'. *Theoria*. 64 (3): 53–76.

Molefe, M. (2018). African Metaphysics and Religious Ethics. *Filosofia Theoretica; Journal of African Philosophy, Culture and Religions*. 7 (8): 19–37.

Molefe, M. (2020). *African Personhood and Applied Ethics*. Makhanda: AHP Publications.

Murove, M. F. (2004). An African Commitment to Ecological Conservation: The Shona Concepts of *Ukama* and *Ubuntu*. *Mankind Quarterly*. XLV (2): 195–215.

Murove, M. F. (Ed.) (2009). *African Ethics: An Anthology of Comparative and Applied Ethics*. Scottville: University of KwaZulu-Natal Press.

Murove, M. F. (2014). Ubuntu. *DIOGENES*. 59 (3–4): 36–47.

Mweshi, J. (2019). The African Emphasis on Harmonious Relations: Implications for Environmental Ethics and Justice. In, Munamato Chemhuru (Ed.) *African Environmental Ethics: A Critical Reader*. Cham: Springer, 191–204.

Organisation of African Unity (OAU). (1981). *African Charter on Human and People's Rights*. (Banjul Charter). Retrieved from www.refworld.org/docid/3ae6b3630.html (Accessed 11 March 2021).

Ramose, M. B. (1999). *African Philosophy Through Ubuntu*. Harare: Mond Books.

Rawls, J. (1971). *A Theory of Justice*. Cambridge: Harvard University Press.

Regan, T. (1987). The Case for Animal Rights. In, M. W Fox and L.D Mickey (Eds.) *Advances in Animal Welfare Science*. Dordrecht: Springer, 179–189.

Samkange, S. and Samkange, T. M. (1980). *Hunhuism or Ubuntuism: A Zimbabwean Indigenous Political Philosophy*. Harare: Graham Publishing.

Singer, P. (1975). *Animal Liberation*. New York: Early Bird Books.

Tempels, P. (1959). *Bantu Philosophy*. (Trans. Rev. Collin King). USA: HBC Publishing.

5 African Ecofeminist Environmental Justice

5.1 Introduction

The nature of African feminist environmental ethical thinking and justice has not been given serious consideration in most philosophical works on African environmental ethics. This, notwithstanding the fact that, elsewhere, it is becoming increasingly clear that "a solution to the questions of environmental justice and environmental ethics needs to start from an understanding of the social relations underpinning current patterns of unsustainability together with an understanding of the material relations between humanity and nature" (Mellor, 2000: 108). According to this view, a conception of environmental justice ought to take into account, or at least be informed by a conception of social justice issues in society in general. Although a lot has been written to this point on the nature of African environmental ethics (Oruka, 1997; Murove, 2004; Kelbessa, 2005; Ojomo, 2010, 2011; Horsthemke, 2015; Chimakonam, 2018; Chemhuru 2019), not much literature exists that focuses specifically on African ecofeminist environmentalism. I therefore seek to examine how African ecofeminist environmental ethical ideas, especially those on the role of women, children and disadvantaged communities, might provide a plausible framework for promoting environmental justice in Africa.

One might wonder what an African ecofeminist environmental justice perspective ought to look like, and whether such a perspective could be considered a plausible alternative to other environmental ethical perspectives in traditional environmentalism. In this chapter, I interpret and advance what I see as ecofeminist views in African philosophies of existence such as communitarian philosophy and *unhu/ubuntu*. I take these philosophies as reasonable premises on which to ground what I see as an alternative ecofeminist environmental ethical thinking in traditional anthropocentric philosophies and traditions. Although I do not claim to be advancing an all-encompassing view through African ecofeminist dimension, I provide various reasons why African communitarian thinking and *unhu/ubuntu* ought to be viewed as capable of informing sound ecofeminist environmental ethical thinking in African philosophy.

DOI: 10.4324/9781003176718-6

Like I have partly alluded to earlier in the first chapter, in terms of documentation, the area of African environmental ethics is still fairly new in African philosophy, having taken shape in the African post-colonial era. According to Horsthemke, "until recently very little has been written on the subject" of African ethics in general (Horsthemke, 2015: 1). Moreover, the area of African ecofeminist environmental ethics itself has not yet received much attention from African environmental philosophers. With the exception of Tangwa (2004), Murove (2004, 2009), Ojomo (2010, 2011), Behrens (2010, 2014) and Horsthemke (2015), who are focused on grounding various underexplored moral and relational ethical theories in African environmental ethics (Chemhuru, 2016: 9), very little mention has been made and very little emphasis has been placed on African ecofeminism as alternative to anthropocentric environmentalism in Africa. Recently, Konik has attempted to bring *unhu/ubuntu* into dialogue with ecofeminism (Konik, 2018). However, her transversal approach to the philosophy of *ubuntu* does not take into consideration the inseparability of the philosophy of *unhu/ubuntu* with African communitarian existence. In contrast to the aforementioned important contributions to African environmental ethics, my claim is that African communitarian philosophy and *unhu/ubuntu* are plausible grounds for constructing African ecofeminist environmental ethics and justice. I therefore venture into this critical and underexplored discourse with a view to examining whether and how African feminist thinking implicit in both communitarian thinking and *unhu/ubuntu* could be plausibly interpreted and adopted as alternatives to anthropocentric environmental ethical thinking and justice in Africa.

I interpret the African philosophies of communitarian thinking and *unhu/ ubuntu* from an African ecofeminist philosophical perspective by gleaning non-anthropocentric elements in the two ontological value systems. As I bring these two African philosophies of existence into conversation with ecofeminist thinking, I argue that both communitarianism and *unhu/ubuntu* could be interpreted as informing a plausible conception of African ecofeminist environmental justice. Such an alternative view could contribute to the growing body of literature on African environmental ethics. As I advance an environmental justice perspective based on the African ecofeminist view, I do not wish to be interpreted as prescribing a complete solution to the African environmental crisis. However, I will argue that African ecofeminist environmentalism could also be considered to be a possible alternative to traditional anthropocentric thinking that is characteristic of traditional philosophical thinking.

I divide this chapter into three sections. First, I consider traditional anthropocentric environmental ethical views that are characteristic of both Western and African philosophical traditions, especially those that are mainly influenced by the Judeo-Christian heritage, Platonic and Aristotelian thinking, philosophical and scientific realism as well as some radical and unrestricted interpretations of African communitarianism and *unhu/ubuntu*. I argue that these approaches to environmental ethics are always in conflict with the quest for environmental justice. In the second section, I present an outline of the social, political and philosophical perspectives that anchor African ecofeminist philosophical thinking in general. I attempt to connect these African ecofeminist philosophical perspectives to the African quest for both social and environmental justice. Having done that, I proceed, in the third and last section, to reconsider how both African communitarian philosophy and *unhu/ubuntu* might inform a plausible African ecofeminist and environmental justice perspective. I divide this section into two sub-sections, one focusing on the ecofeminist import in communitarian philosophy and the other on *unhu/ubuntu* respectively. Generally speaking, I espouse an African ecofeminist view of environmental justice grounded in African communitarian philosophy and *unhu/ubuntu*. I consider it to be a reasonable alternative to the anthropocentric environmental ethical thinking that characterises traditional philosophy.

5.2 Anthropocentric Environmental Ethics as a Threat to Environmental Justice

In this section, I examine some of the major premises of anthropocentric environmental thinking that pose a threat to environmental justice in both the Western and African philosophical traditions. My initial focus on Western philosophical perspectives is intended to show the impact of Western influence on non-Western traditions such as the African philosophical tradition. I show that even if the African tradition has its own indigenous environmental ethical outlook, this has also been greatly shaped and influenced by some anthropocentric perspectives from Western tradition in various ways. Although I later on appeal to the African perspectives in constructing environmental justice based on communitarian philosophy and *unhu/ubuntu*, I also identify some anthropocentric slants in these respective African philosophical traditions. These anthropocentric cavities could be understood and taken to be possible objections to African communitarianism and *unhu/ubuntu*. This is despite my appeal to these philosophies as grounds for conceptualising environmental justice in sections to follow.

Until the works and influence of Lynn White (1967), Aldo Leopold (1949), J. Baird Callicott (1989, 2001) and Holmes Rolston (2000) in the twentieth century, the dominant view in environmental ethics, attributable to the effect of traditional Western philosophical perspectives, could largely be described as anthropocentric. This is mainly due to the influence of various factors such as the Judeo-Christian heritage, Platonic and Aristotelian thinking and philosophical and scientific realism. Misconceptions and misinterpretations of the philosophies of communitarian thinking and *unhu/ubuntu* in African philosophical thinking could also be taken as confirming anthropocentric thinking in the African philosophical tradition. However, if the philosophies of communitarian thinking and *unhu/ubuntu* are conceptualised with due consideration, they could actually contribute to sound environmental ethics and justice, as I will later establish. This is why Godfrey Tangwa juxta-poses Western and African environmental ethical perspectives and comes to the conclusion that the former is "predominantly anthropocentric and indi-vidualistic, and contrasted with its African counterpart, which is described as eco-bio-communitarian" (Tangwa, 2004: 392). Notwithstanding some objections, which I will highlight first in this section, the latter perspective in Tangwa's view is what I seek to show in terms of its import for environmental justice in African philosophy.

Despite the tension between Western and African perspectives on environ-mentalism, for one reason or another, human beings have always generally taken the environment, both physical and non-physical, to be *their* environ-ment, which they need to explore and exploit in order to fulfil their life pur-poses. For that reason, traditionally environmental ethical thinking has been greatly shaped by anthropocentric thinking. By anthropocentric thinking, I mean a human-centred view of ethical considerability. Elsewhere, I have argued that "anthropocentric thinking is the prioritisation of human interests at the expense of everything else" (Chemhuru, 2016: 16). This thinking is based on the assumption that human beings have, and ought to continue to have, dominion and influence over all other beings and non-animate beings in the world. The environmental ethical implications of such thinking are the relegation of the environment and future generations in all matters relat-ing to ethical consideration whenever human beings come into contact with various aspects of the environment. In a highly influential article titled *The Historical Roots of Our Ecological Crisis*, Lynn White (1967) confirms the above view that traditional thinking is inherently anthropocentric, meaning that human beings have generally accepted the human-centred view and approach to ethical thinking. This view has historically shaped the better part of environmental ethical thinking in many African and non-African traditions. However, in the last few decades, there has been a growing body of literature and movements advancing non-anthropocentric environmental-ism such as the 'Land Ethic', the 'Deep Ecology Movement' and the 'Animal Rights Movement' and African philosophies placing emphasis on communi-tarian existence and *unhu/ubuntu*. Although these movements are important in both Western and African environmental ethical traditions, I will not focus my attention on them because they are not part of my discussion here. I will rather, focus more on African communitarianism and *unhu/ubuntu*, although of course there might be some instances where I engage with some of the fundamental positions of these environmental ethical perspectives.

Although there have been multiple influences on anthropocentric perspec-tives, the Judeo-Christian heritage has been held as having strongly influ-enced both African and non-African traditions. Although it might be based on controvertible accusations, the Judeo-Christian heritage view, which is based on the assumption that human beings are superior to all other creatures on planet Earth, has been mainly accused of anchoring traditional anthropo-centric thinking in many of Western and even non-Western traditions such as the African context. According to White, "especially in its Western form, Christianity is the most anthropocentric religion the world has ever seen" (White, 1967: 4). Blames Judeo-Christian attitudes to nature for propagating this kind of anthropocentric thinking, White further argues that the *historical roots of our ecological crisis* stem from Christian attitudes towards nature. For him, "the Christian dogma of creation, which is found in the first Creeds, has another meaning for our comprehension of today's ecological crisis" (White, 1967: 4). Jonathan Kangwa takes White's view further and argues that "the patriarchal understanding that the Bible places man above woman and non-human forms of life has contributed to the marginalisation of women and the natural world" (Kangwa, 2020: 78). However, notwithstanding these anthropocentric interpretations of Judeo-Christian influence on anthropo-centric environmental ethics, a judicious interpretation of Christianity might be that it propagates a non-anthropocentric view of environmentalism. This is because in Judeo-Christian teachings, one mainly finds the ethics of love, care, respect and responsibility, all of which could inform non-anthropocentric environmental ethics and justice.

Apart from the above view on the influence of Judeo-Christian thinking on traditional anthropocentric thinking with reference to African anthropocen-tric perspectives, Western and non-Western philosophical anthropocentric thinking is also thought to be mainly informed and influenced by traditional Platonic and Aristotelian philosophy alike. Platonic and Aristotelean phi-losophy is interpreted to be largely eudaemonist in orientation. Such a view seems to espouse the anthropocentric view of environmental ethical think-ing because of its implicit emphasis on *eudemonia* (happiness or flourishing) which is usually measured with reference to human beings. For that reason, both Plato (427BC–347BC) and Aristotle (384BC–322BC) seem to espouse an anthropocentric view of environmental ethics that privileges humanity over nature simply by virtue of humanity's possession of the faculty of reason. To confirm this anthropocentric view, Aristotle suggests a highly anthropo-centric and hierarchy-based view of environmentalism. For him, "we may infer that after the birth of animals, plants exist for their sake, and that the other animals exist for the sake of man, the tame for use and food, the wild, if not all, at least the greater part of them for food, and for the provision of clothing and various instruments" (Aristotle, 2001, 1256b: 15–22). This hierarchy-based anthropocentric thinking is also implicit in, and taken up in Medieval philosophy by, St. Augustine (354–430) and St. Thomas Aquinas (1225–1274). Aquinas looks at the nature of inequality among beings in the hierarchy of existence and comes to the conclusion that "in natural things species seem to be arranged in degrees; as the mixed things are more perfect than the elements, and plants than minerals, and animals than plants, and men than other animals; and in each of these, one species is more perfect than others" (Aquinas, 1912/2007: 247). This view confirms the anthropocentric thinking that is thought to characterise most theistic traditions. However, one objection that is often given in response to such a view is that, rather, these theistic traditions actually foster a culture of living in accordance with virtues such as peace, love, honesty, harmony and empathy, all of which might be in sync with environmental justice. Notwithstanding this objection however, it cannot be doubted that these theistic views largely support anthropocentric hierarchies of existence among beings.

In addition to the above-mentioned hierarchy-based views of existence, which could be interpreted as authenticating anthropocentric thinking, there is also philosophical and scientific realism in modern philosophy. These philosophical approaches are mostly influenced by thinkers such as Nicolaus Copernicus (1472–1543), Francis Bacon (1561–1626), Isaac Newton (1642–1726) and Galileo Galilei (1564–1642), who could be taken as having been responsible for ushering in the scientific revolution. Over and above their influence on the development of independent thinking and science, their views also strengthen the anthropocentric view of reality. Common among these early modern thinkers is their acceptance of science as the leading tool in obtaining knowledge about the universe. As Henry Odera Oruka observes, "Francis Bacon's philosophy of nature is one of the foundations of the culture of 'modern' Western science and technology. In such a philosophy, modernity and development are seen purely as the continuous domination and utilisation of nature to benefit humankind and humankind alone" (Oruka, 1997: 246). This kind of approach in Bacon and other modern thinkers mainly influenced human beings into thinking that human beings are indeed at the centre of the universe such that they can understand it from an anthropocentric standpoint. As a result, this approach gave rise to the first industrial revolution in Europe in the mid-eightieth century and its subsequent spread to the rest of the world, including in Africa in the ninetieth and twentieth centuries respectively. Such developments have also meant a change of attitude in humanity's conception of reality to a more anthropocentric perspective than before, a view which Oruka sums up as a humans waging a "technological war against nature" (Oruka, 1997: 246). Owing to the success of technological advancements, it became clear that human beings are actually at the centre of the universe and rationally superior to all other beings in the universe. It is therefore not surprising that after the nineteenth and twentieth centuries, environmental problems such as pollution, climate change, global warming and extinction of species and biodiversity began to be felt on a large scale.

Apart from the above anthropocentric conceptions in Western tradition, mention must be made of the African philosophical tradition. Although it cannot be denied that the African philosophies of communitarian thinking and *unhu/ubuntu* point to strong views of communalistic and humane existence, there are some elements of anthropocentrism that cannot be ignored in the two philosophies. Since I do not wish to romanticise African philosophies of communitarian thinking and *unhu/ubuntu*, I will admit and highlight that they are also not immune to accusations of also being anthropocentric. This is because of their main focus is on the existence of human communities and humane existence (mostly among human beings), on which they place emphasis, making indirect reference to environmental ethical concerns. By way of example, a hierarchy-based anthropocentric view of environmental ethical thinking similar to that in Aristotle and Aquinas can also be gleaned from the African ontological hierarchy of existence that is characteristic of African philosophy (Tempels, 1959: 58 and Teffo and Roux, 1998: 138). As I have already shown in Chapters 3 and 4, respectively, African ontology is based on an ontological hierarchy of beings (Magesa, 1997; Ikuenobe 2014). According to this hierarchy, existence is understood in so far as beings are connected from the Supreme Being (God), the ancestors, human beings, animals, down to non-animate beings (Chemhuru, 2016: 104–5). This order is arranged according to the level of potency and could therefore also be interpreted to be anthropocentric in its orientation and import to environmental ethics and justice. It appears that in such a hierarchy, spiritual beings and human beings are morally privileged over non-human beings and the natural environment at large because they occupy the apex of and that human beings depend largely on the other lower beings for their survival and well-being. Accordingly, the same ontological hierarchy of beings can also be criticised for being based on an anthropocentric pyramid such as that espoused by Aristotle, St. Augustine and Thomas Aquinas. Within such a hierarchy, one might wonder how the so-called lower beings such as future generations and non-human animals might benefit equally from the distribution of environmental benefits and burdens. This question will, however, be answered in the sixth chapter.

The above-mentioned threats to environmental justice and the objections which might be raised against African communitarianism and *unhu/ubuntu* in particular, ought to be acknowledged. Nevertheless, I will argue that both African communitarian thinking and *ubuntu* should not be understood from a radical standpoint, where it would appear as if the community is a total imposition over the individual person/s. Different readings of both African communitarian philosophy and *unhu/ubuntu* such as that by Mangena (2009: 18) have tended to portray these two philosophies incompatible with feminist philosophical thinking and environmental justice. According to Mangena, for example, "the African woman's moral point of view is still far from being respected because of the whims and caprices of patriarchy which is camouflaged in the communitarian philosophy of *hunhu* or *ubuntu*" (Mangena, 2009: 18). However, I take Mangena's view as mainly influenced by Western feminist thinking because of the way it overstates the social and political import of African communitarianism and *unhu/ubuntu*. Western feminist thinking based on the dichotomisation of humans on the basis of gender categories is rather thought to be foreign to Africa. In addition, such an approach is believed to be a strategy that is used "for exploiting people and stratifying society" (Oyewumi, 2004: 1). However, this should not be taken to imply that feminism per se in typically un-African, or that African philosophy is essentially against feminist perspectives. My view is that, there are some aspects in Western feminism that are not typical of African feminism. This is why I take a different reading of African communitarian philosophy and *unhu/ubuntu* and argue against certain Western misconceptions about them, while at the same time fleshing out the African ecofeminist philosophical perspectives in the following section.

5.3 The African Ecofeminist Philosophical Perspective

In this section, I focus on the nature of African ecofeminist philosophical thinking and its import for conceptions of environmental justice. First, I examine some of the central premises on which ecofeminist philosophical thinking is grounded as a broader philosophical view. (Warren, 2000; Mellor, 2000). I then situate similar perspectives within the African philosophical context that confirm that "there is a link between the oppression of women and the exploitation of the environment" (Kangwa, 2020: 77). A similar perspective is also seen in, for example, Ojomo (2010), Siwila (2014) and Konik (2018). I consider these and other ecofeminist philosophical perspectives as able to provide a comprehensive view of African environmental justice. I note that in Africa, women, children, black people, the poor and nature should be understood as having suffered the same kind of injustice resulting from traditional patriarchal, anthropocentric and exploitative thinking based on lack of "hospitality, compassion, magnanimity and care for the other" (Nagel, 2013: 181). Hence, my intention to critically explore a plausible view of environmental justice stemming from some of the most influential conceptions of existence, i.e. African communitarianism and *unhu/ubuntu,* which emphasise on communal and humane values, respectively, all of which could at least be useful to a conception of ecofeminist environmental justice.

Broadly speaking, an ecofeminist view is concerned with understanding environmental justice issues by looking at the nature of an 'Othered' social structure. By way of example, Warren refers to unjustified dominated groups such as women, children, people of colour and the poor people as 'human Others' and the environment as consisting of 'earth Others' such as animals, forests, water sources and the land (Warren, 2000: 1). In the same way, I see the hermeneutics of an African ecofeminist philosophy as part of this broader perspective concerned with addressing social and environmental injustice facing both *human others* and *earth others.* Within the African environmental ethical context, I make these connections between social injustice facing othered humans and injustice facing the environment because "the relationship between humanity and nature is heavily circumscribed by relations between human and human" (Mellor, 2000: 108). Yet the dichotomy of the *Self* and the *Other* is quite foreign to the African context because gender and racial categories in Africa became more pronounced during and after the eras of slavery, colonialism and industrialisation in Africa (Oyewumi, 2004: 1). This explains why, for example, the African people have carried the overwhelming burden of environmental injustice associated with slavery,

racism, oppression and colonialism through land dispossession, land degradation, lack of access to natural resources, climate change, global warming
and pollution. However, the effects of these have largely been felt by women,
children and the poor because "women are severely affected than men due
to their social roles as carers and provisioners and their social locations as
the poorest and most vulnerable at the bottom of social hierarchy, alongside children" (Kangwa, 2020: 76). So, a meaningful environmental justice
framework is one that should take into account these realities, especially the
conditions of women and children, which could be similar to that of nature
in various ways.

The domination and oppression of both human beings and nature constitute a central issue in African and other non-African environmental ethical traditions. As I will always emphasise throughout this work, in Africa,
for example, slavery and colonialism have played a significant role in both
human oppression and environmental injustice. Taken together, these twin
processes and the other factors examined in the previous section associated
with anthropocentric thinking and environmental injustice remain the main
culprits of both human oppression and environmental oppression respectively. Accordingly, ecofeminist philosophy should attempt to understand
these philosophies of domination in human societies and their intercon-
nection with the domination of nature because there appears to be a close
relationship between the two. According to Philomena Ojomo, "ecofeminism as a school of thought in environmental ethics seeks to end all forms of
oppression, including that of the environment. It does so by highlighting the
interconnections between the domination of humans by fellow humans on
the basis of race, gender and class on the one hand, and human domination
of the earth on the other" (Ojomo, 2010: 54). From this argument, Ojomo's
view could be interpreted and understood to mean that issues of racism,
colonialism, gender oppression, environmental oppression and injustice in
Africa are ecofeminist issues that all have implications for environmental justice. This is quite a reasonable argument since it is apparent that problems
of injustice in Africa mainly affecting women, black people, poor people
and people with disabilities, children and nature are similar; and that they
can be addressed using almost a similar framework. As Warren sees it, "ecological feminists (ecofeminists) claim that there are important connections
between the unjustified dominations of women, people of colour, children,
and the poor and the unjustified domination of nature" (Warren, 2000: 1).
The African ecofeminist philosophical view therefore challenges all forms of
domination and oppression, especially patriarchy (Kangwa, 2020). It holds
that if the origin of human social and political problems is clearly understood
and addressed, then it will be easier to comprehend and address environmental problems as well. So, ultimately, the way to concretise environmental
justice would be to use a similar approach.

It must be stated that African ecofeminist philosophy sees some of these
philosophies of domination, exploitation, separatism and male chauvinism as

responsible for the suffering, subjugation and exploitation of African philosophy and epistemology. Partly, this is one of the reasons why "much of what we have done in the contemporary history of African philosophy appears to be only corrective work – that is, to respond to bad philosophy that came out of equally bad scholarship on Africa by European social scientists" (Masolo, 2018: 54). Indeed, African philosophy, African ethics, African women, African children, the disadvantaged African people, the African poor people, black African people in general and the African environment at large also suffer the same fate. For that reason, African feminist philosophical perspectives try to integrate these problems in order to have a sort of common and universal approach to them, rather than considering them as independent from each other.

So far, it is clear that problems of oppression and marginalisation are interlinked. This view is partly captured by Kangwa who notes that "women oppression is interwoven with the marginalisation of the environment" (Kangwa, 2020: 78). The same philosophies of domination are largely responsible for the exploitation of the environment and resultant problems such as deforestation, desertification, drought, climate change, poverty, biodiversity crisis, wildlife extinction and the ill treatment of animals and environmental injustice at large. (Warren, 2000: 1.) Accordingly, African ecofeminist philosophy tries to address the same social and political problems resulting from such oppression and injustice by looking at the social and political structures responsible for the same injustices. In this respect, African ecofeminist philosophy becomes a useful approach to justice, fairness and equality between and among human communities, including future generations and the surrounding natural environment. It tries as much as possible to understand environmental justice issues by making connections between social and political concerns on the one hand, and environmental ethical concerns on the other.

The central argument of African ecofeminist thinking is that social and political philosophies, such as patriarchal systems based on the oppression of women, colonialism and dualist thinking are responsible for supporting philosophies of domination, exploitation and colonialism in Africa. As Geraldine Moane puts it, "hierarchy is a central feature of the social context" (Moane, 1999: 24). Similarly, the same epistemologies that are responsible for injustice in society are also thought to be responsible for the traditional division of society and reality into men/women, black/white, humans/nature. Likewise, oppressive and exploitative thinking is also believed not only to support the anthropocentric view of environmentalism, but that it is responsible for environmental injustice in society. From an African social and political perspective, the way oppression and colonisation of African people and their environment have been perpetrated in Africa was based on similar traditional patriarchal, philosophical traditions of domination, oppression and division of the societies that justified colonialism.

If the environmental crisis and injustice facing communities in Africa are to be confronted, various social and political issues relating to human oppression based on unnecessary hierarchisation of society first need to be addressed. Once these social and political hierarchy issues are addressed, it would be easier to comprehend the environmental crisis and perhaps reach a plausible conception of environmental justice. According to Moane, "to understand fully the implications of hierarchical systems … it is first necessary to consider the characteristics of hierarchical systems in greater detail" (Moane, 1999: 24). Accordingly, following Moane's view, addressing the roots of patriarchy, oppression, exploitation, dualist thinking resulting from traditional philosophies of domination in society in general, could go a long way in the search for equality, fairness and justice, all of which are important in solving the various environmental problems facing the world. The thinking is that, an unequal, unfair and unjust society typifies injustice at large, including environmental injustice.

Broadly speaking, the African ecofeminist argument for environmental justice is centred on trying to understand environmental ethical issues from a feminist standpoint. It also involves understanding where African philosophy is coming from as a traditionally suppressed and disadvantaged discourse seeking legitimacy and authentic liberation, just as various aspects of the environment do. African ecofeminist environmental philosophy should therefore play a leading role in connecting the story of African philosophy with that of the natural environment in a traditionally, and inherently anthropocentric context. Once this has been done, African ecofeminist environmentalism could be taken as a reasonable alternative to anthropocentric environmentalism. The conclusion to be drawn is that all environmental ethical problems afflicting humanity cascade down to the oppressed, who are mostly women, suffering because of their traditional 'place' in society. According to Warren, it is important to note that "among white people, people of colour, poor people, children, the elderly, colonized peoples, so-called Third World people, and other human groups harmed by environmental destruction, it is often women who suffer disproportionately higher risks and harms than men" (Warren, 2000: 2). This view therefore challenges humanity to consider the environment as a serious feminist issue. Indeed, if social and problems associated with feminism together with environmental ethical concerns are not addressed, both women and nature will continue to suffer. I therefore adopt this perspective as I proceed in the next section to examine how this African ecofeminist philosophical perspective can be understood as informing sound ecofeminist thinking following the two African philosophies of communitarian thinking and *unhu/ubuntu*.

5.4 African Communitarian Philosophy and *Unhu/Ubuntu*: the Ecofeminist Import

In the previous section, I considered some ecofeminist perspectives which could be read from an African philosophical perspective. In this section, I seek to specifically interpret some of the African philosophies of communitarian thinking and *unhu/buntu* from ecofeminist perspectives. I do so in order see how these respective philosophies might help to construct plausible views of environmental justice in African philosophy. In Chapters 3 and 4, I outlined the nature of environmental justice based on African communitarian philosophy and *unhu/ubuntu* when I examined the African land ethic and African relational environmental justice based on these two philosophies. I will therefore not spend time on what African communitarian philosophy or *unhu/ubuntu* is concerned with per se, or how they are generally oriented towards environmental ethics and justice because I have already done this. What I now seek to do is to flesh out their ecofeminist inclinations and proceed to show how that orientation might be plausibly taken as the grounding for African environmental justice conceptions. I will generally argue that these conceptions of existence form the grounds for plausible communitarian and *unhu/ubuntu*-based ecofeminist environmentalism in African communities. Although both philosophies are inextricably connected to each other, I will examine them separately by beginning with the communitarian view.

5.4.1 The Communitarian Ecofeminist Dimension

In this particular section, I analyse the African communitarian view in terms of its ecofeminist dimension and orientation towards environmental justice. I examine some distinctive features of African communitarian societies such as the matrilineality, anthropomorphism and the feminisation of nature. I consider these in terms of their significance from an ecofeminist perspective to environmental justice. Notwithstanding some conceivably patriarchal features in African communitarian philosophy, a view which Oyewumi (2004: 5) thinks is still objectionable, I focus mainly on its matrilineal and ecofeminist perspectives and seek to interpret some conceptions of ecofeminist environmental justice from it. I examine how these structures of communitarian existence ought to be understood as informing African ecofeminist conceptions of environmental justice.

African communitarian philosophy, especially in its radical view (Menkiti, 1984), might be read as implicitly authenticating an essentially patriarchal structure of community. For this reason, it is viewed as being inherently anthropocentric. However, Gyekye is of the opinion that "the dichotomisation of the individual into patrilineal and matrilineal categories hardly makes sense; there is really no rational or moral justification for it" (Gyekye, 1997: 98). In line with Gyekye's view, I argue that African communitarian societies had a thoroughgoing, non-gendered and somewhat monist view of humans and relations with nature prior to their interface with Western cultures. Oyewumi affirms this view when she observes that "the male/ female, man/woman duality and its attendant male privileging in Western gender categories is particularly alien to many African cultures" (Oyewumi, 2004: 7). Besides looking at this duality as alien to Africa, Oyewumi initially gives the example of the Yoruba among whom "kingship roles and categories are not gender-differentiated" (Oyewumi, 2004: 5). However, this view might be objectionable in certain other African communities in sub-Saharan Africa, where kingship structures are based mainly on patrilineal structures. Nevertheless, some African communitarian societies are largely matrilineal (Siwila, 2014: 135) despite some patrilineal and patriarchal characteristics. By matrilineal, I mean that the same societies are inseparably anchored in the maternal lineage. Although I maintain that most African communitarian societies are inherently matrilineal, I do not wish to be misunderstood here as implying that such communities are matriarchal. I must highlight that matriarchy is different from matrilineality in so far as the former is concerned with a social arrangement characterised by gender categories and female dominance, while the latter has more to do with maternal lineage. This is what makes me consider African communitarian communities as matrilineal, although some patrilineal and patriarchal elements may be evident in the same communities.

The African communitarian structure and its ecofeminist philosophy of environmental justice are incomplete without the African matrilineal structure. As I argue here, this matrilineal structure of African communitarian societies plays a very significant role in inculcating the kind of African ecofeminist environmental justice that I espouse in this chapter. Because of the matrilineal orientation of African communitarian communities, the link between the African communitarian structure and the African matrilineal structure has a very strong ecofeminist dimension that has not been given serious consideration. Although African communitarian ecofeminist philosophical thinking is not entirely new, "there is limited literature on the value of indigenous knowledge to African ecofeminism" (Siwila, 2014: 132) and environmental justice. This is why the matrilineal structure of African communitarian communities needs to be critically interrogated in terms of its environmental ethical import.

Among most African communitarian societies, the role that ecofeminist philosophical thinking could play towards informing sound African indigenous environmental justice cannot be underestimated. In such communitarian and matrilineal communities, for example, indigenous knowledge about environmental justice for the community and the surrounding environment is passed down from one generation to another. Matrilineal structures play a central ecofeminist role in passing down such environmental philosophy and conceptions of justice. I further discuss this view in the sixth chapter as part of the notion of intergenerational environmental justice in African philosophy. According to Siwila, "a grandmother's presence created an opportunity for girls to observe how indigenous knowledge is passed from women to the young girls and how this informed the way in which women would respond to feminist spirituality" (Siwila, 2014: 135). However, Siwila looks at the role of such matrilineal structures in African communitarian societies and limits their ecofeminist environmental ethical import to women and girls alone (Siwila, 2014: 135). This kind of approach is problematic as it could be read as authenticating patriarchy. I will argue that such African indigenous knowledge from matrilineal structures is targeted at all children regardless of their gender. It therefore ought to be taken as a basis for sound ecofeminist environmental justice because ecofeminist environmental knowledge within the matrilineal structure is handed down to all children from generation to generation regardless of their gender.

Following the African matrilineal structure, one is likely to find very strong connections between the communitarian matrilineal structure and ecofeminist environmental justice. It cannot be denied that almost all environmental ethical problems and environmental injustice issues affecting African communities tend to affect women the worst (Warren, 2000; Kangwa 2020). Nevertheless, because of their experience with such environmental problems and environmental injustice, women play a central role in inculcating environmental ethical thinking and environmental justice conceptions (Siwila, 2014). According to a 2015 report by the African Woman Activists (WoMin), "African women, alongside our working-class, indigenous, peasant and black sisters in other parts of the world, offer the most revolutionary alternatives to [the] deeply destructive model of development. These alternatives are found in the ways African women produce food, conserve and steward natural resources and take care of our families and communities" (African Women Activists (WoMin Report), 2015). According to this view, in as much as women stand to be the most affected by environmental policies and practices that are unjust, they tend to positively contribute to environmental well-being and natural resource conservation for the good of their respective communities and future generations.

In African matrilineal and communitarian societies, women, especially grandmothers, aunts, mothers and sisters are the ones who spend most of their time with young children at home, and ultimately fend for their day to day needs of families. As a result, much of the traditional knowledge particularly that relating to the ethics of relational living, respect, empathy, living well and caring for one another, all come from these people through matrilineal structures. Such acquisition of knowledge and wisdom also has implications for environmental ethical thinking and environmental justice, since these are the kind of virtues that are essential to living well with nature. By way of example, the content of such traditional knowledge and wisdom transmitted through these structures could include, but is not limited to, various aspects of environmental ethical wisdom such a taboos intended to protect water sources, plant species, animals and the environment in general (Chemhuru and Masaka, 2010: 121–133). The kind of knowledge and wisdom that these African communitarian and matrilineal structures impart to the young children (both girls and boys) is focused on grooming the individual into more socially and environmentally responsible persons.

Apart from the ecofeminist environmental ethical import of the matrilineal arrangement of African communitarian societies, another ecofeminist view that can also be gleaned from the African communitarian view is based on anthropomorphising and feminising nature or identifying nature with women. This takes nature into some sort of communitarian relations and accepting it as a 'She;. Siwila looks at this perspective based on maternalistic approaches in order to address environmental issues as an ecomaternalistic theory of environmental justice (Siwila, 2014: 135). Although such a view is not peculiar to African ecofeminist environmentalism alone, it sees nature as identifiable with the *Woman* or the *Girl child*. This explains why nature can thus be referred to in feminine terms as *She* or *Mother Nature*. One could interpret this view as a plausible conception of environmental justice because it challenges human beings to bring into solidarity, and consider traditionally oppressed beings together. On the other hand, the idea of feminising nature could be interpreted as being essentially anthropocentric because of its social construction and universalisation of gender, which is essentially a problem (Oyewumi, 2004: 3). Nevertheless, the positive interpretation of this perspective is its identification of women, children and the disadvantaged together with nature so that they can be given attention, respect and justice because of their vulnerability to exploitation. From my earlier argument, it is a fact that nature, women and girl-children have all suffered from patriarchal and exploitative lifestyles of domination through the course of humanity. Accordingly, the attempt to anthropomorphise and feminise nature in African communitarian communities is a quest for environmental justice grounded in the search for social justice. This quest for environmental justice is based on the need to protect all traditionally oppressed beings and is not motivated by any patriarchal or anthropocentric reasons.

In addition, the above view could be reasonably interpreted as having a very strong ecofeminist import to environmental justice based on the honour, respect and reverence accorded to women. This is because African mothers and girl children are honoured, respected and revered because of their fundamental roles that they play in giving and sustaining life in human societies (Mangena, 2009: 26). This is the way nature also ought to be construed, respected and revered because of its role in sustaining human and non-human life. One objection to this view is that it sounds highly anthropocentric. If it is interpreted correctly, however, this perspective is not attached to any anthropocentric reasons, but just to the need to respect nature. As I indicated in Chapter 3 when I looked at the environment from the communitarian land ethic perspective, the honour, respect and reverence it is accorded ought to be understood as emanating from the fact that it sustains life in general. Accordingly, such a view could be acceptable as a life-centred approach to environmental justice, as opposed to an anthropocentric environmental ethical view.

Some feminist thinkers may interpret the ecofeminist dimension from a totally different angle. For Siwila, "this historical ascertaining of women as connected to nature has been problematised by some of the ecofeminist scholars, who argue that the notion could simply re-invent the dualistic wheel of oppression where just as the earth is abused by humanity, so are women"

(Siwila, 2014: 135). However, my response to this kind of objection would be that the African communitarian ecofeminist view of environmental justice is not as fragmented as such a view implies. According to this African communitarian thinking, nature, women, children, people with disabilities and poor people are not only connected to each other, they also form a connection with other human beings, who ultimately have obligations to respect nature, women, people with disabilities and children because they are part of the broader community, although are always disadvantaged.

Generally speaking, it is so far clear that African communitarian philosophy might be plausibly understood as the basis for informing reasonable conceptions of environmental justice. This is because of its ecofeminist emphasis, which is based on the connections between human beings and nature as I have argued here. Apart from this dimension, one other closely connected view to this African communitarian perspective is also the philosophy of *unhu/ubuntu,* which I examine in the following section as also emphasising on environmental justice conceptions based on what it means to be a human being.

5.4.2 The Ecofeminist Import of Unhu/Ubuntu

Having examined the ecofeminist dimension in African communitarian thinking, I now examine the ecofeminist import in *unhu/ubuntu* and see how it might contribute to environmental justice. The philosophy of *unhu/ubuntu* is closely connected to and inseparably anchored in African communitarian thinking. However, there could be some controversy and debate in terms of whether *unhu/ubuntu* is a variant of African communitarianism or whether it is unquestionably communitarian in and for itself (Metz, 2011; Oyowe, 2013). I will not venture into that debate for now, as I treat these terms differently, albeit admitting that they are closely related. As I have already shown in Chapters 1, 3 and 4, *unhu/ubuntu* is generally a Southern African concept of existence that refers to how humaneness is derived from humanness, based on a shared humanity among the various people who are speakers of Bantu languages in sub-Saharan Africa (Samkange and Samkange, 1980; Ramose, 1999). Although the philosophy of *unhu/ubuntu* is sometimes read as a philosophy binding what are thought to be largely patriarchal communities in Africa (Mangena, 2009; Nagel 2013), I interpret its ecofeminist significance to African environmental justice.

A discussion of African communitarian philosophy would not be complete without drawing from *unhu/ubuntu* philosophy. *Unhu/ubuntu* is one of the major defining features of the African communitarian way of life and simultaneously "a promising foundation for human rights" (Metz, 2011: 534) in such communitarian societies. However, the philosophy of *unhu/ubuntu* has often been criticised for propagating 'manly' virtues because of what are seen as embedded patriarchal expectations within it (Mangena, 2009: 18–30; Nagel, 2013: 181). Nevertheless, I will argue to the contrary and look at it differently. By way of example, one of the major premises of the philosophy of *unhu/ubuntu* is that the individual person cannot be understood as being separate from the community, making the person essentially communitarian. The expression, *munhu munhu navanhu* (Shona)/*umuntu ngumuntu ngabantu* (isiNdebele/isiZulu) aptly captures this communitarian element in *ubuntu* philosophy. From this view, Murove thinks that "… the main presumption in *ubuntu* is that the individual is indelibly associated with the community, and can only flourish in *Ukama* [relationship] within the community" (Murove, 2004: 203). As a result, the philosophy of *unhu/ubuntu* must be understood as both a relational and a humanistic philosophy. As a relational view, it places emphasis on relationships between human beings and all other beings, including nature, as I demonstrated in Chapter 4. Although as a humane philosophy of existence, it emphasises the well-being of both human communities and non-human communities that are customarily suppressed and disregarded in much of traditional environmentalism due to the influence of patriarchy. Understood this way, *unhu/ubuntu* therefore becomes a sort of a feminist ethic within its own African communitarian context. Responding to a question about whether *unhu/ubuntu* can be redeemed as a feminist ethic, Nagel contends that "one way it certainly can is to postulate that 'manly' virtue is a deliberate or unconscious biased misapplication of the concept that seems so fundamental to what counts as an African philosophy" (Nagel, 2013: 182). This is why, for example, within African communities that are guided by the philosophy of *unhu/ubuntu*, a good human being is one who treats both human and non-human beings (nature included) in a humane way. This should be an essential feature of *unhu/ubuntu*, because it entails being considerate, kind, magnanimous and compassionate to various aspects of nature that, like human beings, possess life and some of which are sentient. In other words, it is not enough to simply relate well with human beings and mistreat nature. Effectively, *unhu/ubuntu* becomes an ecofeminist ethic responsible for inculcating conceptions of environmental justice among human beings.

As a contribution to environmental ethics and justice, the philosophy of *unhu/ubuntu* closes the supposed gap between and among all beings and species in the African moral world. This is because of its emphasis on the thinking that all human beings are *vanhu/abantu* (human beings) with *unhu/ubuntu* (humaneness). Such kind of human beings are therefore capable of humane existence not only among themselves, but with the rest of the world through relationships. According to Murove, this means that "the existence of *muntu* [human being with *ubuntu*] is in *Ukama* [relationship] with the natural world" (Murove, 2004: 207). This implies that what it means to be, or to exist, is to be the kind of a person who appreciates the interconnectedness between all species that have a relationship with human beings. As a result, *unhu/ ubuntu* challenges philosophies of separatism evident in the way self/other, men/women, humans/nature are thought to be distinct from each other in traditional patriarchal philosophies of oppression and domination that inform strong anthropocentric thinking in much of Western philosophy. The import of this understanding of *ubuntu* to African ecofeminist environmental justice is that within it, there is great emphasis on interrelatedness and interdependence among human beings and nature at large. In terms of environmental justice, such a view might be understood in considering environmental benefits and burdens between different people based on what it means to exist on *unhu/ubuntu*.

Following the kind of thinking explicated above, *unhu/buntu* ecofeminist thinking is therefore incompatible with patriarchal, oppressive, dualist and anthropocentric thinking. This is because such thinking tends to make unnecessary divisions between and among human beings themselves and between human beings and nature at large. For that reason, Ramose looks at "the reductionist, fragmentative and empiricist rationality ... as being responsible for the serious disturbances to ecology" (Ramose, 1999: 157). Accordingly, Ramose thinks that the philosophy of *unhu/ubuntu* needs to be restored in order to strike the right balance in relationships between human beings and nature. For Ramose, "the loss of this balance constitutes a violation of *botho*. It is also an indication of the need to restore *botho* in the sphere of the relations between human beings and physical nature" (Ramose, 1999: 157). Ramose's view can be interpreted to imply that the African philosophy of *unhu/buntu* has an ecofeminist dimension that can be taken as a plausible alternative to anthropocentric thinking. This is because he looks at the relationships between human beings and nature as needing some sort of balance. His view sees nature as being in need of redress through the philosophy of *unhu/buntu*.

Also implicit in the African philosophy of *ubuntu* is the material ecofeminist perspective. Material feminism construes systems such as dualism, capitalism and patriarchy as central to the oppression of women, children, the poor, the disadvantaged and the less privileged. Material ecofeminism is therefore concerned with and more focused on "race, sexuality, imperialism, and anthropocentrism" (Alaimo and Hekman, 2008: 18) and their implications for human-nature relationships. The materialist feminist view is rooted in "corporeal feminism, environmental feminism, and science studies" (Alaimo and Hekman, 2008: 18). It is based on an attempt to confront the material structure of society and the way this might impact on the relationships between human beings and nature (Mellor, 2000: 107). *Unhu/ubuntu* is therefore contrary to materiality because of its emphasis on humaneness among human beings as they relate with each other and with the environment. Similarly, African material ecofeminist thinking in *unhu/ubuntu* is an attempt to understand the ecofeminist argument by looking at the same material or existential conditions shaping humanity's materiality, capitalist and patriarchal frameworks that could be detrimental to sound and non-anthropocentric environmental ethical thinking. Such material ecofeminist thinking implicit in *unhu/ubuntu* therefore brings to the fore, the materiality within the human community, which shapes relationships between human beings and nature. It is based on the view that unless the African material ecofeminist view in *unhu/ubuntu* is understood, human beings will continue to think that traditional anthropocentric cultural and power structures that support the philosophies of dominion, exploitation and inequality are correct. Konik thinks that "material eco-feminism is a version of ecological feminism premised on the assumption that the material conditions of life – economic and environmental ones – shape power relations, economic and cultural practices, skills and ideas" (Konik, 2018: 271). This is why African material ecofeminist views in *unhu/ubuntu* attempt to address existential material conditions such as the materialist, exploitative and anthropocentric thinking. Such existential material conditions are inherent in the traditional view that holds that women and nature are feminine, intrinsically connected, and hence distinct from men.

The import of *unhu/ubuntu* to addressing materiality cannot be overemphasised. Given its emphasis on relational living between and among human beings and other physical and animate aspects of the environment, *unhu/ ubuntu* tries as much as possible to moderate traditional patriarchal, dualist and anthropocentric views of existence. Hapanyengwi-Chemhuru and Shizha, for example, identify some of the values salient in *unhu/ubuntu,* which I see as representative of the ecofeminist thinking in *unhu/ubuntu*. These are "hospitality, fraternity, courtesy, self-sacrifice for the benefit of family and community, kindness, humility, consideration, gentleness, fairness, responsibility, honesty, justice, trustworthiness, hard-work, and integrity and above all tolerance" (Hapanyengwi-Chemhuru and Shizha, 2012: 23). Understood closely from an ecofeminist perspective, *unhu/buntu* environmental ethics emphasises the need to treat various aspects of nature with care, reverence, kindness and accord them ethical consideration. At the same time, following this ecofeminist dimension, virtues derived from *unhu/ubuntu* such as caring, goodness and reverence could also be appealed to in terms of according value to non-animate aspects of nature such as physical nature, plants and water bodies that do not necessarily have sentience.

5.5 Conclusion

In this chapter, I closely examined environmental justice perspectives from indigenous African communitarian communities that are guided by *unhu/ ubuntu*. As I uncovered some ecofeminist perspectives to environmental justice in African communitarianism and *unhu/ubuntu*, I come to almost the same conclusion that Kelbessa draws about indigenous African environmental ethics. For Kelbessa "there is a need, an extremely urgent and ubiquitous need, for the revival of a multiplicity of indigenous and cultural environmental ethics in contemporary societies" (Kelbessa, 2005: 30). Similarly, looking at African communitarian philosophy and *unhu/ubuntu*, I discovered some sound ecofeminist environmental justice perspectives that ought to be seriously considered from African indigenous and cultural communities.

Although Behrens, does not focus so much on African communitarian thinking and *unhu/ubuntu*, he notes that "contrary to anthropocentric ideas, there is a strong emphasis on the inter-relatedness or interconnectedness of human beings with the rest of nature that is evident in African thought, providing basis for a promising African environmentalism" (Behrens, 2014: 65). It is against this background that I attempted to mine some ecofeminist environmental justice conceptions in African communitarian thinking and the philosophy of *unhu/ubuntu*. Overall, I come to the conclusion that these African philosophies could contribute meaningful ecofeminist environmental justice that could possibly replace inherent anthropocentric thinking and environmental injustice in traditional thinking.

References

African Women Activists (WoMin Report) (2015). An African Ecofeminist Perspective on the Paris Climate Change Negotiations. Retrieved from: https://www.peacewomen. org/sites/default/files/An%20African%20Ecofeminist%20P erspective%20on%20the% 20Paris%20Climate%20Negotiations.pdf (Accessed 19 September 2021).

Alaimo, S. and Hekman, S. (2008). Introduction: Emerging Models of Materiality in Feminist Theory. In, Satcy Alaimo and Susan Hekman (Eds.) *Material Feminisms*. Bloomington: Indiana University Press, 1–19.

Aristotle. (2001) Politics. (Trans. Benjamin Jowett.) In, Richard McKeon (Ed.) *The Basic Works of Aristotle*. New York: The Modern Library, 1127–1324.

Aquinas, S. T. (1912/2007). *Summa Theologica: Volume 1, Part 1*. (Trans: Fathers of the English Dominican Province. New York: Cosimo Books.

Behrens, K. (2010). Exploring African Holism With Respect to the Environment. *Environmental Values*. 9 (4): 465–484.

Behrens, K. (2014). An African Relational Environmentalism. *Environmental Ethics*. 36 (1): 63–82.

Callicott, J. B. (1989). *In Defence of the Land Ethic: Essays in Environmental Philosophy*. New York: SUNY Press.

Callicott, J. B. (2001). The Land Ethic. In, Dale Jamieson (Ed.) *A Companion to Environmental Philosophy*. Malden: Blackwell, 204–217.

Chemhuru, M. and Masaka, D. (2010). Taboos as Sources of Shona People's Environmental Ethics. *Journal of Sustainable Development in Africa*. 12 (7): 121–133.

Chemhuru, M. (2016). *The Import of African Ontology for Environmental Ethics*. D. Litt et Phil. (Philosophy) [Unpublished]: University of Johannesburg. Retrieved from: https://ujcontent.uj.ac.za/vital/manager/index?site_name=Research%output (Accessed 1 March 2017).

Chemhuru, M. (Ed.) (2019). *African Environmental Ethics: A Critical Reader*. Cham: Springer Nature.

Chimakonam, J. O. (2018). *African Philosophy and Environmental Conservation*. New York: Routledge.

Gyekye, K. (1997). *Tradition and Modernity: Philosophical Reflections on the African Experience*. Oxford: Oxford University Press.

Hapanyengwi-Chemhuru, O. and Shizha, E. (2012). *Unhu/Ubuntu* and Education for Reconciliation in Zimbabwe. *Journal of Contemporary Issues in Zimbabwe*. 7 (2): 16–27.

Horsthemke, K. (2015). *Animals and African Ethics*. New York: Palgrave-Macmillan.

Ikuenobe, P. A. (2014). Traditional African Environmental Ethics and Colonial Legacy. *International Journal of Philosophy and Theology*. 2 (4): 1–21.

Kangwa, J. (2020). Women and Nature in the Book of Job: An African Ecofeminist Reading. *Feminist Theology*. 29 (1): 75–90.

Kelbessa, W. (2005). The Rehabilitation of Indigenous Environmental Ethics in Africa. *Diogenes*, 207: 17–34.

Konik, I. (2018). *Ubuntu* and Ecofeminism: Value-Building an African and Womanist Voices. *Environmental Values*, 27: 269–288.

Leopold, A. (1949). *A Sand County Almanac and Sketches Here and There*. Oxford: Oxford University Press.

Magesa, L. (1997). *African Religion: The Moral Traditions of Abundant Life*. New York: Orbis Books.

Mangena, F. (2009). The Search for an African Feminist Ethic: A Zimbabwean Perspective. *Journal of International Women's Studies*. 11 (2): 18–30.

Masolo, D. A. (2018). History of Philosophy as a Problem: Our Case. In, Edwin E. Etieyibo (Ed.) *Method, Substance, and the Future of African Philosophy*. Cham: Palgrave Macmillan, 53–69.

Mellor, M. (2000). Feminism and Environmental Ethics: A Materialist Perspective. *Ethics and the Environment*. 5 (1): 107–123.

Metz, T. (2011). Ubuntu as a Moral Theory and Human Rights in South Africa. *African Human Rights Law Journal*, 11: 532–559.

Menkiti, I. A. (1984). Person and Community in African Traditional Thought. In, Richard A. Wright (Ed.) *African Philosophy: An Introduction*. Lanham: University Press of Americas, 171–180.

Moane, G. (1999). *Gender and Colonialism: A Psychological Analysis of Oppression and Liberation*. New York: Palgrave Macmillan.

Murove, M. F. (2004). An African Commitment to Ecological Conservation: The Shona Concept of *Ukama* and *Ubuntu*. *The Mankind Quarterly*. 45 (2): 195–215.

Murove, M. F. (2009). An African Ethics Based on the Concept of *Ukama* and *Ubuntu*. In, Munyaradzi F. Murove (Ed.) *African Ethics: An Anthology of Comparative and Applied Ethics*. Pietermaritzburg: University of KwaZulu-Natal Press, 315–331.

Nagel, M. E. (2013). An Ubuntu Ethics of Punishment. In, Mechthild E. Nagel and Anthony J. Nocella II (Eds.) *The End of Prisons: Reflections from the Decarceration Movement*. Amsterdam – New York: Radopi, 177–187.

Ojomo, P. A. (2010). An African Understanding of Environmental Ethics. *Thought and Practice: A Journal of the Philosophical Association of Kenya*. 2 (2): 49–63.

Ojomo, P. A. (2011). African Environmental Ethics: An African Understanding. *The Journal of Pan African Studies*. 4 (3): 101–113.

Oruka, H. O. (1997). *Practical Philosophy: In Search of an Ethical Minimum*. Nairobi: East African Educational Publishers.

Oyewumi, O. (2004). Conceptualising Gender: Eurocentric Foundations of Feminist Concepts and the Challenge of African Epistemologies. In, Signe Arnfred et al (Eds). *African Gender Scholarship: Concepts, Methodologies and Paradigms*. Dakar: CODESRIA, 1–8.

Oyowe, A. O. (2013). Strange Bedfellows: Rethinking *Ubuntu* and Human Rights in South Africa. *African Human Rights Law Journal*, 13: 103–124.

Ramose, M. B. (1999). *African Philosophy Through Ubuntu*. Harare: Mond Books.

Rolston, H. (2000). The Land Ethic at the Turn of the Millennium. *Biodiversity and Conservation*. 9 (2000): 1045–1058.

Samkange, S. and Samkange, T. M. (1980). *Hunhuism or Ubuntuism: A Zimbabwean Indigenous Political Philosophy*. Salisbury: Graham Publishing.

Siwila, L. C. (2014). "Tracing the Ecological Footprints of Our Foremothers": Towards an African Ecofeminist Approach to Women's Connectedness With Nature. *Studia Historiae Ecclesiasticae*. 40 (2): 137–147.

Tangwa, G. B. (2004). Some African Reflection on Biomedical and Environmental Ethics. In, Kwasi Wiredu (Ed.) *A Companion to African Philosophy*. Malden: Blackwell Publishers, 387–395.

Teffo, L. A. and Roux, A. P. J. (1998). Metaphysical Thinking in Africa. In, Peter H. Coetzee and Abraham P. J. Roux (Eds.) *Philosophy in Africa: A Text With Readings*. Johannesburg: International Thomson Publishing Southern Africa, 134–148.

Tempels, P. (1959). *Bantu Philosophy* (Trans. Reverend Colin King). Paris: Présence Africaine.

Warren, K. J. (2000). *Ecofeminist Philosophy: A Western Perspective on What It Is and Why It Matters*. Lanham: Rawman and Littlefield Publishers.

White, L. (1967). The Historical Roots of Our Ecological Crisis. [With Discussion of St. Francis; Reprint, 1967.]. In, David Spring and Eileen Spring (Eds.) *Ecology and Religion in History*. New York: Harper and Row, 1–7.

6 Intergenerational Environmental Justice in African Philosophy

6.1 Introduction

What, if anything do human beings owe to future generations of both human and non-human beings? This question raises a fundamental problem of the feasibility of *serving justice to different generations* comprising of both human beings and non-human beings. The answer to the problem of distributive intergenerational justice lies in the extent to which "the present generation is bound to respect the claims of its successors" (Rawls, 1971: 251), that consist of both human and non-human beings. However, it should be overemphasised that the solution to this problem is not easy to arrive at. As John Rawls acknowledges, this problem "subjects any ethical theory to severe if not impossible tests" (Rawls, 1971: 251). Notwithstanding the challenges associated with reference to a theory of intergenerational environmental justice (IEJ) as seen by Rawls, intergenerational concerns remain central to any theory of environmental justice. In this particular chapter, I appeal to the African communitarian model in order to advance a novel grounding from which to approach IEJ that takes into account the claims of future communities.

In the growing body of literature on environmental ethics, the question of whether current generations have any ethical obligations towards future generations has been fairly well addressed. This could be due, in part, to responses to the increasing threat to the natural environment as well as the need to safeguard justice as one of what Rawls sees as the uncompromising first virtues of society (Rawls, 1971: 4). Although "increasing attention has been paid to the question of intergenerational distributive justice" (Beckerman, 1997: 392), scholarship has not been exhaustive, since some fundamental questions of IEJ have remained underexplored. Yet, this fundamental concern for the long-term health and prosperity of future generations (Golub, Maren and Harlow, 2013: 274) remains important. Essentially, the notion of IEJ ought to deal with and address the problem of who ought to be involved in the distribution of environmental benefits and burdens resulting from human action on the environment. I will examine this understanding of environmental justice in greater detail below when I make reference to various conceptions of it such as those by Robert Bullard (1990), the American Environmental Protection

DOI: 10.4324/9781003176718-7

Agency (2008) and the United Nations Development Programme (2014). These conceptions, I observe, have largely influenced some conceptions of environmental justice in non-Western traditions such as African philosophy. In the same vein, despite the growing body of literature on African environmental ethics and justice lately (see, for example, Murove, 2004, 2009; Behrens, 2012; Ikuenobe, 2014; Kelbessa, 2015; Chemhuru, 2016, 2019a, 2019b), the area of IEJ has not received much attention from African environmental philosophers.

In this chapter, I seek to make a fresh contribution to the less explored debate on IEJ in African philosophy. I do so by attempting to flesh out some of the ontological, teleological and ethical contributions of the African communitarian view of existence in terms of their import to intergenerational equity. I argue from a rather uncontroversial premise about African communities, namely that such communities are inherently communitarian. I have already examined the meaning of this term in great detail in previous chapters with reference to the views of John Mbiti, Ifeanyi Menkiti and Kwame Gyekye who have popularised such a view, at least in academic writing. What I do in this particular chapter is to analyse the import of such an Afro-communitarian philosophy to IEJ and proffer reasons why the appeal to Afro-communitarian philosophy could be viewed as a viable approach to distributive IEJ in Africa. In this way, I see my theory as capable of contributing to a plausible, intellectually robust and practical conception of IEJ within an African communitarian context.

This chapter is arranged as follows: First, I consider some of the generally accepted traditional conceptions of environmental justice, generally in terms of the fact that they do not explicitly offer a comprehensive conception of IEJ. I consider how they focus mostly on equity among the current generations while they are silent about future generations. I seek to provide ethical arguments in favour of IEJ. In the second section, I seek to achieve an understanding and appreciation of the ethics of IEJ. This I do by providing a philosophical justification of the notion of IEJ. In the third section, I locate the notion of IEJ within the African environmental ethical context. I consider the question of whether African environmental ethics might have a conception of intergenerational environmental ethics. I then demonstrate how this question can be addressed in the affirmative in the last section by appealing to the African conceptions of existence, and the African communitarian view of existence in particular. Overall, I attempt to present the African communitarian view of existence as the basis for my argument for IEJ.

6.2 The Limits of Some Traditional Conceptions of Environmental Justice

In this section, I focus on some of the influential conceptions of environmental justice so far, in terms of how they might be helpful in constructing conceptions of IEJ. I first note that it is important to have a conception of what environmental justice is before attempting to get into the more complex issue of IEJ in African philosophy. Although it is generally agreed among scholars that "there is no single universally accepted perspective on environmental justice issues" (Danielson, 2019: 212), the understanding of IEJ is actually informed and determined by how one understands environmental justice. It is for this reason that I first revisit some of the traditional and most influential conceptions of environmental justice in this section notwithstanding the diversity of approaches in such discourse. I do this in order to show how such conceptions, albeit largely significant, are somewhat vague and largely lacking on the aspect of IEJ that I am concerned with.

As I have already shown in Chapter 2, environmental justice is generally construed as being a late twentieth century discourse in environmental politics. It is aimed primarily at addressing "unpersuasive global environmental policies and practices that support inequitable distribution of environmental benefits and burdens" (Chemhuru, 2019a: 32). According to this view, environmental justice is mainly concerned with the quest for the equitable sharing and distribution of the benefits and burdens associated with environmental use, depletion of natural resources or climate change between people of different communities, classes, race, gender, environments and generations. However, ideas about how to involve all generations including future generations seem to have been conspicuously absent from much of the discourse on environmental justice in the twentieth century in general. Bullard (1990), for example, published one of the first and most influential books on environmental justice, *Dumping in Dixie: Race, Class and Environmental Quality.* This book focuses mainly on the race-based environmental inequalities and politics in the United States of America, especially among "blacks and other minorities, the poor and working class persons" (Bullard, 1990: 1). From this book, it is clear that the understanding of environmental injustice based on racial politics is an important aspect of environmental justice discourse the world over. However, questions of environmental justice and injustice based on intergenerational obligations are equally important but conspicuously absent from much of this earlier discourse.

Apart from Bullard, the other most influential view of environmental justice has been provided by the American Environmental Protection Agency (EPA). As I have already shown in Chapter 2, the EPA has proffered the following widely accepted view of environmental justice:

Environmental Justice is the fair treatment and meaningful involvement of all people regardless of race, colour, national origin, culture, education or income with respect to the development, implementation and enforcement of environmental laws, regulations and policies. Fair Treatment means that no group of people, including racial, ethnic, or socioeconomic groups, should bear a disproportionate share of the negative environmental consequences resulting from industrial, municipal and commercial operations or the execution of federal, state, local and tribal environmental programs and policies. Meaningful Involvement means that: (1) potentially affected community residents have an appropriate opportunity to participate in decisions about a proposed activity that will affect their environment and/or health; (2) the public's contribution can influence the regulatory agency's decision; (3) the concerns of all participants involved will be considered in the decision-making process; and (4) the decision makers seek out and facilitate the involvement of those potentially affected

(Environmental Protection Agency (EPA), 2008)

This view of environmental justice by the EPA is useful but somewhat insufficient for the conceptualisation of IEJ. Firstly, this understanding of environmental justice does not take into account, and is silent about the fate of future generations in terms of their involvement and considerability in the distribution of environmental benefits and burdens. For that reason, given the fact that environmental benefits and burdens in particular, transcend generations, one would have expected that by using the term *meaningful involvement*, the above view of environmental justice would have been explicit about involving future generations. Secondly, the EPA's conception of the *potentially affected* also does not take future generations into account. It seems to limit the *potentially affected* to the present generations only, rendering it to be a limited view of environmental justice. Notwithstanding these observations, however, the EPA's understanding of environmental justice remains one of the most influential view in environmental justice debate.

In addition to the views of Bullard (1990) and the EPA (2008), in 2014 the United Nations Development Programme provided a view of environmental justice that I would consider to be closer to a conception of IEJ. According to the UNDP, "at its core, environmental justice is about legal transformations aimed at curbing abuses of power that result in the poor and vulnerable suffering disproportionate impacts in pollution and lacking equal opportunity to access and benefit from natural resources" (UNDP, 2014: 17). This view could perhaps be considerately interpreted as placing more emphasis on the need to take consideration of the most *vulnerable* and *poor* communities, a view which cannot be disputed for its nobility. However, as I understand it properly, the reference *vulnerable* and *poor* implies that the concerns of these disadvantaged are actually presently living human communities. Most likely these are human beings because it is unusual to describe non-human beings as *poor*, thus rendering such a view limited as far as IEJ is concerned. IEJ concerns ought to be broader than the interests of human beings and current generations. On the other hand, by making reference to individuals *lacking equal opportunity to access,* the UNDP implies that the concerns about future communities ought to be considered in the distributive patterns of environmental benefits and burdens, although of course such a view is still oriented towards human interests. Of exception however, are the UNDP's seventeen Sustainable Development Goals (SDGs) for human development set to be achieved by 2030. However, although sustainable development is a very important ingredient of intergenerational justice, these SDGs could be understood as largely oriented towards human sustainable development, with less emphasis on non-human beings. For example, out of the seventeen development goals, only SDGs number 12 and 13, explicitly address issues of "responsible consumption, and production" (UNDP, 2021: SDG 12), and "climate action" (UNDP, 2021: SDG 13). The rest of the other SDGs could be considered to be largely aggressive development-oriented goals in favour of humanity. Nevertheless, all the SDGs could be taken as useful towards the understanding of IEJ because sustainable development and intergenerational justice are difficult to separate.

Recently, US president, Joseph Biden has tried to show some commitment towards understanding and being involved in environmental justice issues, as opposed to his predecessor, Donald Trump, who appeared reluctant to tackle environmental injustice outside the American context. Soon after inauguration on 20 January 2021, Biden quickly signed an executive order that sought to address environmental justice and climate change issues and made the following declaration:

> Our Nation has an abiding commitment to empower our workers and communities; promote and protect our public health and the environment; and conserve our national treasures and monuments, places that secure our national memory. Where the Federal Government has failed to meet that commitment in the past, it must advance environmental justice. In carrying out this charge, the Federal Government must be guided by the best science and be protected by processes that ensure the integrity of Federal decision-making. It is, therefore, the policy of my Administration to listen to the science; to improve public health and protect our environment; to ensure access to clean air and water; to limit exposure to dangerous chemicals and pesticides; to hold polluters accountable, including those who disproportionately harm communities of color and low-income communities; to reduce greenhouse gas emissions; to bolster resilience to the impacts of climate change; to restore and expand our national treasures and monuments; and to prioritize both environmental justice and the creation of the well-paying union jobs necessary to deliver on these goals
>
> (Biden, 2021: January 20).

This declaration was significant as it signalled a departure from the previous administration's stance on climate and environmental justice issues. However noble it may be, this understanding and commitment to environmental justice issues is largely applicable and relative to the American context, thereby leaning on to what some would look at as environmental racism, a perspective that I would avoid to get into detail in this work. This view is clearly seen from the emphasis in the use of "Our Nation". This, despite the fact that climate change and environmental justice issues transcend national, geographic and generational boundaries. As such, approaches to these issues also should be holistic by taking into consideration everyone. As Kelbessa rightly notes, "climate change has intra and intergenerational justice dimensions because it places both present and future human and non-human generations into jeopardy" (Kelbessa, 2015: 61). This assertion shows that the obligations towards environmental justice seem to be mainly informed by the spectre of environmental crisis and climate change that currently threatens the whole world. Yet, such obligations for IEJ ought to be broadly informed by the interrelationships and needs of current and future generations.

Also of concern from the above executive order by Boden is the fact that focus is placed on issues mostly affecting current generations such as human public health, clean water and air, reducing climate change by minimising greenhouse gas emissions and ensuring the equal treatment of people of colour and those from low income groups in issues relating to environmental policy-making. However, Kelbessa reminds us that "climate change raises issues of corrective justice and intergenerational justice, as emissions of greenhouse gases negatively affect not only the current generations of humankind but also future ones, non-human species and the natural environment because they can stay in the atmosphere for hundreds and even thousands of years" (Kelbessa, 2015: 46). Although these issues are central to any conception of environmental justice, Biden's declaration on climate change and environmental justice is silent on how IEJ could be conceptualised. It does not clearly show the obligations that the current generations might have towards future generations as well as how to safeguard the environment in the interests of these future generations. Nevertheless, this is not surprising because countries of the global North continue to explicitly ignore ethical and justice considerations in their climate or environmental policies.

In general terms, if justice is to be acceptable as simply fairness (Rawls, 1971: 3) regardless of one's age, race, gender, class and generation among others, then it must also be applicable to environmental justice issues. There is need to critically engage this discourse in the light of more pressing issues relating to how best environmental benefits and burdens could be distributed across generations without necessarily imposing costs on future generations for the benefits of current generations (McShane, 2009: 413). As I will establish, a view of environmental justice that is implicit within the African communitarian philosophy tries to accommodate this need for equity between generations in the distribution of environmental benefits and burdens. However, before I do that, I consider, in the following section, how the general conceptions of environmental justice ought to be transcendentally understood by focusing on IEJ and providing a defence for it.

6.3 The Ethics of Intergenerational Environmental Justice (IEJ)

IEJ deals with the problem of equity with reference to possibilities of claiming, distributing and balancing environmental benefits and burdens among present and future generations. It is an extended view of environmental justice that takes into consideration the interests, needs, entitlements and claims of different generations, including future ones. It is interesting to note that IEJ takes into consideration, the interests of human and non-human beings that are not yet there, and who might also not exist, depending on our actions. Such beings "nevertheless have interests that generate obligations for those of us currently alive" (McShane, 2009: 413). One of the reasons for accepting such a view is that the actions that present living human beings do are capable of affecting the interests of those human and non-human beings to come. For this reason, present generation ought to have duties towards future generations. As Otto Spijkers puts it, "there are various ideas about what exactly the present people owe to future people, and that is what intergenerational equity is all about" (Spijkers, 2018: 1). Indeed IEJ essentially deals with issues such as the question of whether current generations owe anything to future generations in terms of environmental ethical benefits or burdens. What environmental ethical obligations do present generations have towards future generations? Where, do such obligations to future generations, if they exist, come from? What is the basis for defending an egalitarian view of IEJ between people of different generations? In this section, I address these questions as I provide an ethical defence of IEJ.

The quest and justification for equitably distributing environmental justice among different generations remains problematic. It is not only a challenge to environmental theorists and political philosophers, but to policy-makers as well. This explains Richard Hiskes' scepticism about environmental ethical arguments for environmental protection, policy-making and sustainability (Hiskes, 2005: 177). Indeed, questions of equity relating to environmental justice are difficult to address. This is because they could be reasonably approached from various egalitarian positions and theories ranging from Aristotelian ethics, Kantian normative ethical perspectives, utilitarian perspectives and the social contact approaches, amongst others. Responses to issues of IEJ also depend on a number of factors that determine the circumstances of justice. For example, equality for Aristotle should be understood as the distribution of certain goods to people who are equal in merit (Aristotle, 2001: 1280a, 10–25). For Aristotle, therefore, it is possible that there may be inequalities among different human beings. From such a view, it is therefore possible that present generations may prejudice future generations of environmental equity on the basis that beings belonging to different generations may not be equal.

The above view from Aristotle might also seem to be accepted by John Rawls, who also thinks that the 'circumstances of justice', or the conditions subsisting between different generations, the social minimum will also determine "how far the present generation is bound to respect the claims of its successors" (Rawls, 1971: 251). However, while Rawls could be read as also confirming a somewhat self-conceited and relativist view of IEJ among generations, he actually supports an egalitarian view of intergenerational equity. Although Rawls does not articulate IEJ as such, he provides a systematic argument for thinking about distributing justice between generations, which I find to be useful to my defence of IEJ within an African communitarian context. This is why, in thinking of a reasonable justification for an ethical defence of an IEJ framework, I therefore first appeal to Rawls' two principles of justice. I consider them in terms of how they tend to prescribe certain contractarian rights, duties and obligations to individuals in different social, political and economic conditions. According to this contractual scheme, it is possible to envisage the possibility of justice as fairness and equitably distribute it following the two principles of justice as set out by Rawls as follows:

> Each person is to have an equal right to the most extensive scheme of equal liberties compatible with a similar scheme of liberties for others. Social and economic inequalities are to be arranged so that they are both: a). reasonably expected to be to everyone's advantage, and b). attached to positions and offices open to all
>
> (Rawls, 1971: 53).

These principles of justice are therefore central to my ethical justification of a distributive IEJ framework. This is because of their emphasis on some of the intrinsic values such as justice, liberty, welfare, integrity, compassion and equality. These values are intrinsically good and are important for the well-being of both present and future generations. From these principles of justice, Rawls seems to defend equal rights to access, equality for all, concern for the liberties of others and concern for social and economic inequalities among human beings. Rawls' reference to *others*, *everyone* and *all* could be understood to mean both the present and future generations. One of the objections that is often raised against this view is that we cannot talk of either rights or obligations for 'persons' that are not yet born because "persons not yet living cannot be said to have rights" (Hiskes, 2005: 178). Nevertheless, the basis for the ethical considerability of these future generations as denoted by *others*, *everyone* and *all* is that they are potential moral agents whose liberty, welfare, integrity could either be better or worse off.

For Rawls, too, the principles of justice are supposed to apply "to the basic structure of society and govern the assignment of rights and duties and regulate the distribution of social and economic advantages" (Rawls, 1971: 53). In this regard, the level of social minimum for IEJ is simply set by reference to the intrinsic goods already alluded to in the two principles of justice. These intrinsic goods are justice, liberty, welfare, integrity, compassion and equality. By social minimum, I am referring to the framework designed to ensure fairness in distributing environmental benefits and burdens, i.e. among different generations. In other words, the social minimum can be taken to be the minimum conditions that are needed for any meaningful conception of social justice. In African philosophy, a view that comes close to Rawls' idea of the social minimum is Oruka's conception of the right to a human minimum (Oruka, 1997: 81), although he discusses it in a slightly different context, mainly focusing on human poverty, human rights, justice and politics. Nevertheless, Oruka's view could be read as exclusively oriented towards human beings, and thus, purely anthropocentric if it is applied to the context of environmental justice. Accordingly, with reference to IEJ, Rawls's view would be more appealing than Oruka's view because of its broadness in the conceptualisation of intergenerational justice. Rawls, for example, thinks that it will depend on the extent to which the current generation is ethically obliged to respect the rights of the future (Rawls, 1971: 251), which he implicitly believes to be the case. This is because if justice is fairness (Rawls, 1971: 3–4, 8), then "fairness demands ignoring the temporary location of particular generations" (Heyd, 2008: 177). This is why I take the Rawlsian conception of intergenerational justice as plausibly forming the basis for a strong argument in favour of IEJ as opposed to Oruka's anthropocentric-oriented view.

Although Rawls is not very explicit about IEJ, I take his framework for conceptualising 'justice between generations' to be an ethical defence of an IEJ perspective. This is because of his emphasis on the need to take into consideration 'fair shares' in the distribution of benefits and burdens between different generations. According to Rawls, "the just savings principle can be regarded as an understanding between generations to carry their fair share of burden of realising and preserving a just society" (Rawls, 1971: 257). In this way, Rawls' *just savings principle* can be regarded as a largely communitarian and ethical social contract between different generations for the need to have fair shares with reference to the environment. For him, "in following the just savings principle, each generation makes a contribution to those coming later and receives from its predecessor" (Rawls, 1971: 254). This view is similar to the contractarian outlook of African communitarian philosophy. It lays emphasis on interconnections between the various beings in the hierarchy of African communitarian existence. In it, respect for communitarian existence is intergenerational in so far as it is not limited to a particular time-specific community or generation.

From the above intergenerational ethical import in Rawls' theory of justice, one can see that it is compatible with African communitarian philosophy. It is therefore the reason why I take it together with the African communitarian view in order to conceptualise a view of IEJ. The threat to the environment owing to unsustainable use of nature's resources and "the increasing concern with environmental hazards" (Beckerman, 1997: 392) demand that human beings rethink about their present conditions and the future in terms of intergenerational distributive environmental justice.

A similar notion to this idea, although it is not synonymous with IEJ, is that of sustainable development. Sustainable development is the kind of development where human beings are supposed to satisfy their present and immediate needs without compromising those of future generations. According to Kola Odeku, this entails having "policies, development plans, green activists and judicial decisions that look beyond the welfare of the present generations, by ensuring the utilisation of land, water, forest, wildlife, minerals and air resources for the interests of present and future generations" (Odeku, 2012: 185). While IEJ and sustainable development are somewhat interconnected, I take the position that sustainable development is an important ingredient of IEJ. In order for sustainable development to be achieved, there ought to be sufficient conditions for IEJ. Beckerman sums up this view by arguing that "the moral claim of 'sustainable development' is alleged to rest on its appeal to intergenerational equity" (Beckerman, 1997: 397). According to this view, human beings therefore have an ethical obligation to worry about their their welfare and that of the future generations. Consequently, if human beings take their well-being and development to be important, then they should also be worried about the well-being of future generations. This is why Rawls thinks that, "from a moral point of view, there are no grounds for discounting future well-being on the basis of pure time preference ..." (Rawls, 1971: 253). Given the need for sustainable development, human beings therefore have ethical obligations towards future generations because they can either enhance their well-being or harm it. In a way, this could also be understood to be a communitarian view of sustainable development because of its consideration for future communities.

IEJ emphasises that present generations must leave the natural environment and all other resources available to them in a good condition as they have found them, so that future generations/communities can also make use of them. Following this conception of IEJ based on the ethics of sustainable development, my argument for the need for present generations to think about future generations is also premised on the communitarian understanding that present communities have ethical obligations to be considerate and humane with respect to the future communities because they all belong to the same community, in spite of generational gaps and differences. When these obligations are properly understood, it would be possible to defend a plausible conception of IEJ. To confirm this view, Beckerman notes that "most people are distressed by the sight (or knowledge) of the various manifestations of poverty, suffering, sickness and so on. In any given society, it is natural to resent the existence of such conditions" (Beckerman, 1997: 401). Since no one would really want future generations to suffer because of the harmful actions by the present generations towards the environment, IEJ therefore becomes not only a communitarian, humane approach to environmental ethics but also ultimately an ethical pursuit.

By and large, taking into account the above views, I mainly defend intergenerational justice as a noble ethical pursuit. However, an objection that is

often raised with regards to attempts to equitably distribute environmental justice across generations is the difference in terms of time and location
between generations. This view brings in Rawls' dimension of the difference in 'circumstances of justice', which are "the normal conditions under
which human cooperation is both possible and necessary" (Rawls, 1971:
109). According to this view, the circumstance of justice between present and
future generations is not essentially mutual and reciprocal such that relations
with the future do not meet circumstances of justice (Hiskes, 2005: 186; see
also Heyd, 2008: 169). This poses the difficulty of imagining mutuality and
reciprocity between different generations with regard to IEJ. However, this
argument is not convincing because it is not clear whether we ought not to
have ethical obligations towards human beings with whom there are differences in time and location. According to Rawls, "the mere difference of
location in time of something's being earlier or later, is not in itself a rational
ground for having more or less regard for it" (Rawls, 1971: 259). This is
why I appeal to African communitarian philosophy's view of IEJ. I appeal
to it because of its emphasis on taking all human generations (both living
and non-living) and the environment at large as essentially a community
of related beings. In the next section, I will proceed to present the African
communitarian model as an alternative framework for conceptualising IEJ
in Africa.

6.4 Intergenerational Environmental Justice in African Environmental Ethics

In the previous section, I considered what IEJ entails and how to ground it
in environmental philosophy broadly, without necessarily situating it within
any specific philosophical tradition. What I seek to do in this section is to
consider IEJ within the African environmental ethical context. In general
terms, I will briefly look at African environmental ethics as largely oriented
towards intergenerational environmental ethics, especially if African communitarian philosophy is properly interpreted.

As I already established in Chapter 1, it is only after the twenty first century that there has been a growing body of literature on African environmental ethics, including works by Murove (2004, 2009), Horsthemke (2015),
Chimakonam (2018) and Chemhuru (2016, 2019a, 2019b). Notwithstanding
the significance of these works to African environmental ethics broadly, I
maintain that environmental ethical thinking has always been part and parcel of the African indigenous communitarian traditions except that it has
not been clearly documented. Despite this lack of documentation of African
environmental ethics, however, conceptions of existence in African philosophy have always offered a comprehensive environmental ethic from generation to generation, as I established in Chapters 3-5. This is why I now seek
to venture into the African communitarian view of existence and present an
underexplored view of IEJ based on what it means to exist as a community.

In African environmental ethics, quite a considerable body of literature now exists on environmental justice conceptions (see, for example, Chimakonam, 2018; Chemhuru, 2019a; Ssebunya, Morgan and Okyere-Manu, 2019). Despite the existence of this body of literature on African environmental ethics and environmental justice, there still remains a gap in the area of IEJ. With the exception for Kelbessa's (2015) perspective, most of these African environmental justice perspectives have not been concerned with IEJ issues. Hence, to add to this body of works available on African environmental ethics and justice so far, I look at and interpret African communitarian philosophy in order to ground a plausible conception of IEJ that is salient in African environmental ethics.

African environmental ethics could essentially be taken as being concerned with intergenerational environmental ethical issues. It ought to be understood to focus on the import of the African worldviews with reference to the surrounding environment in its entirety. It is a holistic and comprehensive philosophy of existence that deals with fundamental African metaphysical and ethical issues relating to how human beings ought to exist. At the same time, it deals with how human beings can relate well with nature so that they can continue to live well without prejudicing themselves, the environment and the future generations. Capturing this communitarian and intergenerational view of environmental ethical thinking in African philosophy, Polycarp Ikuenobe opines that:

> African views and thought on ontology, cosmology, medicine and health, and religious practices supported their moral attitudes toward the conservation and preservation of nature. Traditional African thought sees nature as holistic and as an interconnected continuum of humans and all natural objects which exist in harmony (Ikuenobe, 2014: 2).

This assertion highlights how issues on African environmental ethics have generally revolved around questions relating to ideas of human existence in relation to the moral status of various beings (humans and non-humans), questions of animal rights, pollution and environmental justice, among others. Most of these issues have generally been approached by appealing to African relational approaches such as *ukama, unhu/ubuntu* and communitarian philosophy (Murove, 2004; Horsthemke, 2015). Although the words *ukama* and *unhu/ubuntu* could be available in languages other than the Shona language in Zimbabwe, these words are used in the Shona language to mean relatedness and humanness, respectively. Such views have generally represented the nature of African communitarian environmental ethics.

What I have done so far is to attempt to generally situate the issues of IEJ within the African environmental ethical context. To achieve a comprehensive understanding of African environmental ethics, I see the need to also mine a perspective of IEJ from within the African communitarian discourse. This would enable African philosophers to think of how African communitarian philosophy serves environmental justice to both the present and future generations. This is why I proceed in the following section to specifically expose the nature of IEJ in African communitarian philosophy.

6.5 African Communitarian Philosophy and Intergenerational Environmental Justice

Within discourse on intergenerational justice, one finds a variety of approaches and ideas relating to the potential sources of ethical obligations to future generations and justification for such appeals. Most of these approaches have mainly been shaped by various appeals to ethical theories in Western philosophy. Here, I seek to appeal to the African view of existence. I focus on African communitarian philosophy in order to advance it as a different framework for thinking about IEJ. Kelbessa has set the tone for such an undertaking as he contends that "African worldviews include intergenerational ethics that teaches that natural resources ought not to be exploited beyond limit, and that the land ought to be taken care of for the benefit of present and future generations, as well as for the good of non-human species" (Kelbessa, 2015: 59). As I pursue the same kind of argument, I seek to focus specifically on African communitarian philosophy and to glean a reasonable conception of IEJ from it.

In the previous chapters, and in Chapters 3 and 4 in particular, I showed that the idea of African communitarian philosophy has mainly been popularised by thinkers such as John Mbiti (1969), Ifeanyi Menkiti (1984, 2004) and Kwame Gyekye (1987, 1992, 2010). Recently, it has been subjected to different interpretations from a variety of scholars. Although African communitarian philosophy is sometimes conflated with the philosophy of *ubuntu*, the two are not the same. However, I avoid getting into a separate discussion of *Unhu/ubuntu* because I think that the normative import of the idea of *ubuntu* is encapsulated within communitarian existence at large. *Unhu/ubuntu* is more of a normative theory of existence that is used with reference to the understanding of an individual primarily amongst Southern African communities, while African communitarian philosophy is broadly a collective view of existence largely defining and characterising African communal existence in general. From this view, *unhu/ubuntu* ought to be taken and understood as merely one important aspect of such African communitarian philosophy. In addition, the notion of *unhu/ubuntu* is not applicable to the rest of Africa, while communitarianism is fairly applicable to the greater part of sub-Saharan Africa. Interestingly, both communitarian philosophy and *unhu/ubuntu* point to strong views of communalistic and humanistic existence that could be interpreted as informing a conception of IEJ because of their emphasis on human needs and interests (Gyekye, 2013: 243). My nuanced focus on African communitarianism is motivated by the need to come up with a philosophy that broadly speak to the general African condition rather than one that is limited to a few communities in Africa. Nevertheless, this approach should not be taken to imply that *unhu/ubuntu* is any less important as it also significantly feeds into communitarian conceptions of humaneness based on what it means to be a human being (humanness).

The central thesis of African communitarian philosophy, which is important to IEJ, is its socio-centric view of being. It lays emphasis on the existence and well-being of the whole community rather than focusing on individual existence. From the socio-centric perspective of this view, Menkiti argues that it is the community that "defines a person as a person, not some isolated static quality of rationality, will, or memory" (Menkiti, 1984: 172). Menkiti's view is usually objected to for placing too emphasis on the role of the community in defining the person. Nevertheless, it is useful in understanding the role of the community towards shaping an individual who is conscious of the surrounding community (environment) or reality and not necessarily the self. According to Menkiti, "the African view of man denies that persons can be defined by focusing on this or that physical or psychological characteristic of the lone individual. Rather, man is defined by reference to the environing community" (Menkiti, 1984: 171). This view could be taken as a useful relational understanding of the person notwithstanding its overemphasis on the role of the community over the individual. It is for this reason that I considerately read it together with Gyekye's moderate and compatibilist view of African communitarian philosophy. For Gyekye, communitarianism "… gives accommodation to the communal values as well as to the values of individuality, to social commitments as well as to duties of self-attention" (Gyekye, 1992: 121) because of its amphibious nature (Gyekye, 1987: 31). These communitarian conceptions are useful to my conceptualisation of a just individual, who ought to reasonably contribute to a conception of IEJ because of the duties and obligations that the individual has towards the community and the environment at large. According to Gyekye, "the natural sociality of the individual immediately involves one in some social and moral roles in the form of obligations, commitments, and duties (or, responsibilities) to other members of his or her community which the individual must fulfil" (Gyekye, 2013: 234). This perspective could be read as giving the individual and present communities the duties towards justice for the future communities because of the duties and responsibility that individuals have. The emphasis on individuals having duty and responsibility could be taken to mean that such individuals have some sort of trust, power and control over how they can relate with and treat present and future communities.

African communitarian philosophy implies ethical obligations towards each other not only among present communities, but future communities as well. It does not limit ethical obligations to the present community alone. This philosophy holds that, the present and the future generations are merely different communities that are related and connected to each other by virtue of their communitarian nature of existence. It might be difficult to conceive how the present generation (community), the future generation (community) and the environment could be related and connected, and whether such different generations could form a community with the past, present and future generations such that it might be said to have ethical obligations towards future generations. However, by virtue of being a community, and the fact that the present community is capable of indirectly harming future communities through its actions on the environment, it follows that the present generation ought to have ethical obligations towards future communities and not necessarily towards particular individuals. In other words, such ethical obligations are not directed towards *future individuals* as such, but towards *future communities*. This view is further strengthened by the communitarian axiom that: "whatever happens to the individual happens to the whole group, and whatever happens to the whole group happens to the individual. The individual can only say: *I am because we are; and since we are, therefore I am*" (Mbiti, 1969: 106). The person understands his/her complete existence by reference to others, who may include past, present and future communities. Understood in this way, such a view of existence could be interpreted as guaranteeing collective IEJ because the ethical obligations for IEJ are not derived from the need to protect and guarantee the interests of *future individuals*. Rather, emphasis is on obligations towards *future communities*. This is why for Kelbessa it is important to know that "there is no need to know the exact nature and demands of autonomous individuals, as the focus is on groups, that is, generational communities" (Kelbessa, 2015: 60). The implications of this view are that once the needs and interests of a community are guaranteed, then there is no need to think about individual self-interests. In the same way, because of the communitarian nature of society, it is impossible for present communities to look at themselves as if they are independent of other future communities. Communitarian existence will always remind present generations or communities of the need to always be aware of their relationships with, and the needs of other future communities.

One possible objection that may be raised to this *future communities* conception of IEJ is the question of whether ethical obligations from present generations are derived from 'direct' or 'indirect' duties towards future communities? This is because it is difficult for present generations to directly wrong future generations that are not presently involved or may not even exist at all. For Metz, "the idea of something being the object of a 'direct' duty [is when it is] owed a duty in its own right, or the idea of something that can be wronged" (Metz, 2019: 11). However, future generations cannot possibly be directly wronged by present generations here and now. Yet, the actions on nature by the present human communities can eventually indirectly affect the welfare of future communities. For this reason, present generations should have ethical obligations towards future generations because they have 'indirect' duties to them. Indirect moral duty towards future generations means that current generations of human beings have moral obligations to treat the environment in a certain way "but not because of facts about it", but because they are related to the future generations (Metz, 2019: 11). Although Behrens looks at this view of 'indirect' moral duties towards future communities as "backward-looking", he positively construes it as "one of the most significant contributions African thought can make to our conception of moral obligations towards the future" (Behrens, 2012: 187).

According to African communitarian philosophy, human beings owe it to future generations to bequeath a healthy and liveable environment. This view is premised on communitarian existence, which emphasises the need for moral considerability between present communities, future communities and the environment at large. This is contrary to some non-African traditions, where the ethical obligations towards IEJ tend to be unidirectional in so far as there is no reciprocity and mutuality between different generations. This view is confirmed by Heyd, who argues that "on the intergenerational dimension, dependence seems to be in principle *unidirectional*, i.e. involves relations which are not and will never be based on reciprocity" (Heyd, 2008: 170). This explains why such non-African traditions in the global North are failing to come up with robust conceptions and frameworks for IEJ in their climate and environmental policies, as I have stated earlier in this chapter. However, this is contrary to the African communitarian context, where there is a strong ethical relationship between, for example, present and future generations as well as between the chain of beings within the African hierarchy of existence.

The other basis for IEJ in African communitarian societies is the notion of moral considerability among beings belonging to different generations. Moral considerability between and among generations is rooted in relational ethics, which is characteristic of, and implicit in, African communitarian philosophy. Gyekye observes that African communitarian ethics is largely characterised by "natural sociality and relationality" (Gyekye, 2013: 234). In this particular philosophy, the dominant view is that present generations are related to future generations, and this ultimately determines and shapes conceptions of intergenerational environmental equity. Although one of the objections that has often been raised against the quest for ethical consideration of future generations has been that we do not really know who these future generations will be, African communitarian philosophy emphasises a relational view of existence, where all beings and communities (both present and future) are related, even if we do not know them. Murove captures this view by focusing on the Shona people of Zimbabwe's notion of *ukama* (relatedness) and argues athat "while the Shona word *Ukama* means relatedness, *Ubuntu* implies that humanness is derived from our relatedness with others, not only those currently living, but also through the generations, past and future" (Murove, 2004: 196). *Unhu/ubuntu* and *ukama* are mostly understood as emphasising relationships and respect among all beings even if such beings are not necessarily related to each other. As such, similar relational ethics in African communitarian philosophy based on *unhu/ubuntu* and *ukama* could be understood to be inclusive of past and future generations because of its humanistic orientation (Gyekye, 2013: 234) and its transcendental view of relationships among beings.

Within African communitarian societies, the community is not under-stood as just a composite of individuals forming the present and static gen-eration of individuals. In other word, community is not only active in the present sense only. Rather, the understanding of community is so complex that it takes into account the obligations of individuals to the present com-munity and future communities as well. Although Menkiti (1984: 173) is not explicit about this intergenerational perspective, he aptly captures the max-imal understanding of the person, where personhood ought to be acquired as a requisite for communitarian existence. This, despite the fact that "per-sonhood is something at which individuals could fail, at which they could be competent or ineffective, better or worse" (Menkiti, 1984: 173). It is for this reason that existence is mainly characterised by African ontological connec-tions of beings with other beings, or communities with other communities because "persons are the sort of entities that are owed the duties of justice" (Menkiti, 1984: 177). In this case, even if Menkiti's concept of the person excludes infants, by implication, future persons/communities are also owed such duties of justice for as long as they will develop from an *it* into a person. Similarly, although infants and future communities might initially be taken as lacking the basis for moral considerability because of their *it* status, the fact that they have the potential to develop from that *it* to some future commu-nity or beings means that they ought to be accorded moral considerability.

The web of existence in African communitarian ontology also defines the nature of African intergenerational environmental philosophy. Since com-munitarian existence is defined by communal relations, or relational exist-ence, it is such that all beings in the web of existence are interconnected to form what Kevin Behrens calls a "moral circle" (2012: 182). This moral circle includes past, present and future beings, each of which beings has connec-tions with and obligations towards the others. Ikuenobe sums up this view in the following passage:

> Traditional African views of ontology can be understood in terms of their view of cosmology. Reality is seen as a composite, unity and harmony of natural forces. Reality is a holistic community of mutually reinforcing natural life forces consisting of human communities (families, villages, nations, and humanity), spirits, gods, deities, stones, sand, mountains, rivers, plants, and animals. Everything in reality has a vital force or energy such that the harmonious interactions among them strengthen reality. For some African peoples such as the Bantu, things in reality can be placed in the following hierarchically ontological categories, based on their power to strengthen interactions and harmony in reality: God, ancestors or spirits, humans, animals, plants, and non-biological things
> (Ikuenobe, 2014: 2)

In keeping with the above exposé, the chain of beings and life forces that are referred to could also be understood not only to be the Supreme Being, ancestors, humans and nature, but that both present and future generations are being taken into account in terms of moral considerability. Although Ikuenobe is silent about this view, this ontological understanding of environmentalism could be interpreted to imply that environmental justice concerns are not only limited to the present generations alone. They also include past and future generations because these are communally interconnected with present generations. In that regard, Behrens argues that "the interconnectedness of the parts of the web of life transcends generations. Clearly future people cannot reciprocate the efforts made by the present on their behalf, but this does not free the living from moral responsibilities towards the yet to be born, not in this African view, at least" (Behrens, 2012: 182). This vitalist view of existence can be the starting point for a reasonable framework for equitable distribution of environmental benefits and burdens to all generations both present and future. This could perhaps explain why Kelbessa comes to the conclusion that "in many African societies, the members of the clan include the unborn, the living in the world of the ordinary sense experience, and those living in the post-mortem world of ancestors" (Kelbessa, 2015: 59). In that regard, because of its communitarian orientation towards respect for the past generations, the present generations and the unborn (future generations), this could be understood to be a vitalist view of IEJ. In such a view, individuals are always reminded of the interconnectedness between the "currently living human and non-human beings, the living dead, the yet unborn, and the natural world" (Kelbessa, 2015: 59).

The view that there are ontological and teleological connections between the various beings in reality could be understood to be an important aspect of an African communitarian framework for conceptualising IEJ. By way of example, the acceptance that there is a connection that stretches from the Supreme Being, ancestors and down to nature could also be understood as transcending into future generations. Although it is based on a metaphysical premise, this hierarchy is thought to stretch from the Supreme Being, the ancestors (which are past generations), human beings (present generations), physical nature and future generations. This hierarchy-based view of environmental ethical thinking can be gleaned from the African ontological hierarchy of existence (Tempels, 1959: 58; Teffo and Roux, 1998: 138). According to this hierarchy of ontology, existence is understood in so far as beings connect from the Supreme Being (God), the ancestors, human beings, animals, down to non-animate beings (Chemhuru, 2016: 104–5). This order is also arranged according to the level of potency and hence could also be interpreted to be oriented towards intergenerational environmental philosophy. According to this metaphysical view, negative human action on nature ultimately has some impact on either previous or future generations because they share their habitat (the environment). Although some of Tempels' conclusions are likely to be challenged for some of his unfair generalisations about African philosophy, he captures this teleological view of communitarian existence as follows:

After [God] come the first fathers of men, founders of the different clans. These archi-patriarchs were the first to whom God communicated his vital force, with the power of exercising their influences on all posterity. They constitute the most important chain binding men to God. They occupy so exalted a rank in the Bantu [African] thought that they are not regarded merely as ordinary dead

(Tempels, 1959: 61–62).

Wiredu further exposes this view when he argues that, "of all the duties owed to the ancestors, none is more imperious than that of husbanding the resources of the land so as to leave it in good shape for posterity" (Wiredu, 1994: 46). If properly understood, then, this communitarian understanding of existence could be the basis for ontological, teleological and ethical connections between the various beings and the future generations.

African communitarian philosophy also inculcates a communitarian ownership and stewardship of natural resources, which has a strong bearing on IEJ. According to this view, nature in its entirety belongs to the community, i.e. past, present and future. For this reason, the environment is viewed as 'communal property' rather than private property, which could be owned by individuals. (Behrens, 2012: 183). Accordingly, the land, water sources and wildlife all belong to the past, present and future communities. This explains why most African axioms and taboos mostly inculcate environmental ethical teachings that are meant to preserve the environment for its own good and for the good of the present and future generations.

6.6 Conclusion

Arriving at an acceptable view of IEJ might be easier said than done given the difficulties of reasonably grounding such a view. What I have done in this chapter is to present the African communitarian model as an alternative framework for conceptualising IEJ, one based on the metaphysical, ethical and teleological view of existence in African philosophy. I do not want to claim that my appeal to African communitarian philosophy is the only comprehensive, or even the best, model available so far. However, some of its ethical appeals to IEJ seem to be convincing. It is for this reason why I come to the conclusion that African communitarian philosophy could be considered as providing sufficient grounds for IEJ.

References

Aristotle (2001). Politics. (Trans. Benjamin Jowett.) In, Richard MacKeon (Ed.) *The Basic Works of Aristotle.* New York: The Modern Library, 1137–1324.
Beckerman, W. (1997). Debate: Intergenerational Equity and the Environment. *The Journal of Political Philosophy.* 5 (4): 392–405.
Behrens, K. G. (2012). Moral Obligations Towards Future Generations in African Thought. *Journal of Global Ethics.* 18 (2–3): 179–191.

Biden, J. R. (2021). *Executive order on Protecting Public Health and the Environment and Restoring Science to Tackle the Climate crisis*. Retrieved from: https://www.whitehouse.gov/briefing-room/presidential-actions/2021/01/20/executive-order-protecting-public-health-and-environment-and-restoring-science-to-tackle-climate-crisis/. White House: 20 January. Date Downloaded: 6 July 2021.

Bullard, R. D. (1990). *Dumping in Dixie: Race, Class and Environmental Quality*. USA: Westview Press.

Chemhuru, M. (2016). *The Import of African Ontology for Environmental Ethics*. D. Litt et Phil. (Philosophy) [Unpublished]: University of Johannesburg. Retrieved from: https://ujcontent.uj.ac.za/vital/manager/index?site_name=Research%output

Chemhuru, M. (2019a). The Paradox of Global Environmental Justice: Appealing to the Distributive Justice Framework for the Global South. *South African Journal of Philosophy*. 38 (1): 30–39.

Chemhuru, M. (Ed.) (2019b). *African Environmental Ethics: A Critical Reader*. Cham: Springer.

Chimakonam, J. O. (Ed.) (2018). *African Philosophy and Environmental Conservation*. London: Routledge.

Danielson, S. (2019). Accessing Discourse of Environmental Justice in the University Classroom. *Environmental Justice*. 12 (5): 212–217.

Environmental Protection Agency (EPA). (2008). Environmental Justice: Learn about Environmental Justice. http://www.epa.gov/environmentaljustice/learn-about-environmental-justice.

Golub, A., Maren, M. and Harlow, J. (2013). Sustainability and Intergenerational Equity. Do Past Injustices Matter? *Sustainability Science*. 2013 (8): 269–277.

Gyekye, K. (1987). *The Unexamined Life: Philosophy and the African Experience*. (Inaugural Lecture Delivered at the University of Ghana, on 7 May 1987). Accra: Ghana Universities Press.

Gyekye, K. (1992). Person and Community in Akan Thought. In, Kwasi Wiredu and Kwame Gyekye (Eds.) *Person and Community*. Washington D.C: The Council for Research in Values and Philosophy, 101–122.

Gyekye, K. (2010). Person and Community in the Akan Thought. In, Kwasi Wiredu and Kwame Gyekye (Eds.) *Person and Community: Ghanaian Philosophical Studies 1*. Washington D.C: The Council for Research in Values and Philosophy, 101–122.

Gyekye, K. (2013). *Philosophy, Culture and Vision: African Perspectives*. Accra: Sub-Saharan Publishers.

Heyd, D. (2008). A Value or an Obligation? Rawls on Justice to Future Generations. In, Axel Gosseries and Lucas H. Meyer (Eds.) *Intergenerational Justice*. Oxford: Oxford University Press, 170–189.

Hiskes, R. P. (2005). Environmental Rights, Intergenerational Justice, and Reciprocity With the Future. *Public Affairs Quarterly*. 19 (3): 177–194.

Horsthemke, K. (2015). *Animals and African Ethics*. New York: Palgrave Macmillan.

Ikuenobe, P. A. (2014). Traditional African Environmental Ethics and African Legacy. *International Journal of Philosophy and Theology*. 2 (4): 1–21.

Kelbessa, W. (2015). Climate Ethics and Policy in Africa. *Thought and Practice: A Journal of the Philosophical Association of Kenya (PAK)*. 7 (2): 41–84.

Mbiti, J. S. (1969). *African Religions and Philosophy*. London: Heinemann.

McShane, K. (2009). Environmental Ethics: An Overview. *Philosophy Compass*. 4 (3): 407–420.

Menkiti, I. A. (1984). Person and Community in African Traditional Thought. In, Richard A. Wright (Ed.) *African Philosophy: An Introduction*. New York: University Press of America, 171–182.

Menkiti, I. A. (2004). On the Normative Conception of a Person. In, Kwasi Wiredu (Ed.) *A Companion to African Philosophy*. Malden: Blackwell Publishers, 324–331.

Metz T. (2019). An African Theory of Moral Status: A Relational Alternative to Individualism and Holism. In, Munamato Chemhuru (Ed.) *African Environmental Ethics: A Critical Reader*. Cham: Springer, 9–27.

Murove, M. F. (2004). An African Commitment to Ecological Conservation: The Shona Concepts of *Ukama* and *Ubuntu*. *The Mankind Quarterly*. XLV (2): 195–215.

Murove, M. F. (Ed.) (2009). *African Ethics: An Anthology of Comparative and Applied Ethics*. Pietermaritzburg: University of KwaZulu-Natal Press.

Odeku, K. O. (2012). Climate Change and Intergenerational Justice: Perspective from South Africa. *Journal of Human Ecology*. 39 (3): 183–194.

Oruka, H. O. (1997). *Practical Philosophy: In Search of an Ethical Minimum*. Nairobi: East African Educational Publishers.

Rawls, J. (1971). *A Theory of Justice*. Cambridge: Harvard University Press.

Spijkers, O. (2018) Intergenerational Equity and the Sustainable Development Goals. *Sustainability*. 10 (3836): 1–12. Downloaded from: www.mdpi.com/journal/sustainability. DOI: 10.3390/su10113836.

Ssebunya, M., Morgan, S. N. and Okyere-Manu, B. D. (2019). Environmental Justice: Towards and African Perspective. In, Munamato Chemhuru (Ed.) *African Environmental Ethics: A Critical Reader*. Cham: Springer, 175–189.

Teffo, L. A. and Roux, A. P. J. (1998). Metaphysical Thinking in Africa. In, Peter H. Coetzee and Abraham P. J. Roux (Eds.) *Philosophy in Africa: A Text With Readings*. Hohannesburg: International Thomson Publishing Southern Africa, 134–148.

Tempels, P. (1959). *Bantu Philosophy* (Trans. Reverend Colin King). Paris: Présence Africaine.

UNDP Report. (2014). *Environmental Justice: Comparative Experiences in Legal Empowerment*. Retrieved from: https://www.undp.org/content/dam/undp/library/Democratic%20Governance/Access%20to%20Justice%20and%20Rule%20of%20Law/Environmental-Justice-Comparative-Experiences.pdf (Accessed 23 September 2021).

UNDP. (2021). *Sustainable Development Goals*. Retrieved from: https://www1.undp.org/content/oslo-governance-centre/en/home/sustainable-development-goals.html (Accessed 23 September 2021).

Wiredu, K. (1994). Philosophy, Humankind and the Environment. In, Henry Odera Oruka (Ed.) *Philosophy, Humanity and Ecology*. Nairobi: ACTS Press, 30–48.

Index

African Charter on Human and People's
    Rights (ACHPR) 49, 87
African communitarianism 4–5, 8,
    14, 20, 22–23, 25–28, 58, 114, 133;
    African environmental ethics and
    25; African relational ethics in 85;
    approach to environmental justice
    52; based on *unhu/ubuntu* and *ukama*
    75–76, 79–81, 100–102, 105–106,
    133–134, 136; conceptions of land 70;
    intergenerational environmental justice
    and 133–139; Menkiti's concept of
    person 137; social and political thinking
    84; view of existence 70, 77; views on
    personhood 92
African ecofeminist environmental
    justice: communitarian view 110–114;
    forms of domination and oppression
    106–109; in Judeo-Christian teachings
    102–103; materialist feminist view
    116–117; 'Othered' social structure
    106; *unhu/ubuntu* philosophy 99–102,
    104–106, 109–110, 114–117
African environmental ethics 8–10,
    21–28; anthropocentric and non-
    anthropocentric ethics 22; ecospheric
    view 27; environmental justice and
    28–30, 36; historiography of 22;
    metaphysical/ontological conceptions
    22–24; *mitupo* (totems) and *ukama*
    (relatedness) 26; nationalist thinkers
    25; ontological, religious and
    communitarian views 23, 25, 27; taboo
    and totem wisdom 26; teleological and
    vitalist views 24–25; twentieth century
    22; writings 26
African environmental justice 1–6;
    global North *vs* global South 4, 34–35,

41; impact of colonialism on 48; in
    sub-Saharan Africa 35–37, 39–41
African environmental philosophy 10,
    16–21; African religions and 17–18;
    Bantu philosophy 17; historiography
    20; pluriversality view of knowledge
    19; reaction to slavery, racism and
    colonialism 19; sage philosophy or
    philosophic sagacity in 21; of savage 17;
    universalist perspective 18–19; writings
    17
African land ethic 10–11, 55–56;
    African ontology-based view 56,
    63–69; 'allocations' or distributions
    61; biocentric view 59; colonial
    61–63; commoditisation of land 71–72;
    communitarian view of 70, 72; in
    environmental justice framework
    57–60, 69–72; expression *son/daughter of
    the soil* 64; Gyekye's view 70; historical
    perspective 60–63; idea of property
    in 69–72; individual ownership of
    property 72; pre-colonial era 61; *unhu/
    ubuntu* philosophy 67–68
African ontology 24, 56–57, 61, 63–64,
    67–69, 82–83, 85, 90, 105; Gyekye's
    view 63; Ikuenobe's views 23, 81,
    137–138; *see also* ontology-based
    African land ethic
African philosophy 20–22; ancient 9; *see
    also unhu/ubuntu* philosophy
African Philosophy Through Ubuntu
    (Ramose) 26
African relational environmental ethics
    4–5, 75–76; agent-related partiality
    89; attainment of *beingness* or fullness
    87; communitarian perspective 84–85;
    conception of 76–82; distinction

For Product Safety Concerns and Information please contact our EU
representative GPSR@taylorandfrancis.com
Taylor & Francis Verlag GmbH, Kaufingerstraße 24, 80331 München, Germany

www.ingramcontent.com/pod-product-compliance
Lightning Source LLC
Chambersburg PA
CBHW060316220326
41598CB00027B/4345